The SEARCH for FOOD

Text by Marco Ferrari

Illustrations by Ivan Stalio

RAINTREE STECK-VAUGHN PUBLISHERS
A Steck-Vaughn Company
Austin, Texas

Published by Raintree Steck-Vaughn Publishers, an imprint of Steck-Vaughn Company

Consultant: Gregory Haenel, Ph. D., Rutgers University
Editor: Kathy DeVico
Electronic Production: Lyda Guz, Scott Melcer
Project Manager: Joyce Spicer

Library of Congress Cataloging-in-Publication Data
Ferrari, Marco, 1954-
The search for food / text by Marco Ferrari; illustrated by Ivan Stalio.
p. cm. — (Everyday life of animals)
Includes bibliographical references (p. 64) and index.
Summary: Discusses the different foods that various animals eat and how they go about getting what they need to survive.
ISBN 0-8172-4195-7
1. Animals—Food—Juvenile literature. [1. Animals—Food habits.] I. Stalio, Ivan, ill. II. Title. III. Series.
QL756.5.F47 1999
591.5'3 — dc21 98-16703
CIP AC

Printed in Italy by Grafiche Editoriali Padane, Cremona
Bound in the United States

1 2 3 4 5 6 7 8 9 0 02 01 00 99 98

Photo credits [and Acknowledgments] that appear on page 64 constitute an extension of this copyright page.

Contents

Introduction

A cheetah in hot pursuit of a gazelle, a wolf pack closing in on a deer, a spider lying in wait for a fly, and a monkey cramming berries into its mouth are all players in a drama that takes place every day in the animal world—the unending search for food.

All animals have to eat in order to get energy for living. They eat plants or other animals, and each kind of animal has its favorite food and its own special way of obtaining it. Over millions of years, the processes of **evolution** have created a huge variety of fascinating methods of catching food. Some meat-eaters, or carnivores, undertake long-distance chases, while others grab their prey with lightning-fast strikes. Some lie in wait for unwary victims, and some use bait to attract them. A few carnivores, including many spiders, even make traps to snare their prey. Wolves hunt in packs, with each member having a particular task and knowing just what to do. Almost all living creatures can be another animal's dinner, but animals don't just sit around waiting to be eaten. Most of them have some kind of defense: some escape simply because they are faster than their enemies; some protect themselves with spines, horns, or other weapons; and others employ **camouflage**—colors and patterns that blend in with the surroundings and make the animals difficult to see. Some are actually poisonous. **Predators** have to be fast or very clever to catch their food. Even plants can protect themselves: many have spines or poisonous juices that deter herbivores (plant-eating animals). But even the best defenses cannot protect against all enemies. Many animals have tough jaws that break down the defenses, or special digestive juices that destroy the poisons.

The giraffe's long neck and tongue are special adaptations for reaching the topmost leaves on the acacia tree.

An animal's **metabolism** determines how much food it needs. **Reptiles** and **amphibians** can make do with less food than the more active birds and **mammals** of the same size.

The Search for Food takes a look at the wide variety of methods animals use to find and capture their food, and also at the many ways in which they avoid being eaten themselves. It presents a fascinating picture of the complex **food chains** and webs that link animals and plants together and keep nature in balance.

An alligator lurks in the water (opposite).

Survival of the Fittest

During the millions of years that animals have been on the Earth, they have had to compete with each other for food. Many species lost the struggle and died out. Only the fittest or most efficient kinds of animals have survived. Most have adapted to a particular type of food, so the Earth's food resources are efficiently divided, and all the animals normally get enough to eat. Most animals are either plant-eaters (herbivores) or meat-eaters (carnivores). In any **ecosystem**, there are enough plant-eaters to feed the meat-eaters, but not so many that they eat all the plants. If the plant-eaters get too common, they will destroy the plants and then starve unless they can find food somewhere else. The meat-eaters then have to move or starve as well.

Each animal species has evolved efficient senses for finding its food. It also has effective equipment for catching or collecting it and the right processes for digesting the food and making the best use of it. The teeth of meat-eating mammals, for example, are very different from the teeth of plant-eaters. In addition to having to find food, animals have to avoid being eaten, so they try to stay one step ahead of their enemies. For example, if a big cat evolves into a faster creature, it will be able to catch more prey. But then prey will evolve to become even faster, restoring the balance. If this did not happen, the prey would die out, and the predator could not eat.

FINDING FOOD IN PALM TREES
The coconut crab, or robber crab, lives on tropical beaches. Unlike most other crabs, it spends most of its time on land. It lives in burrows and holes in trees, sometimes quite far from the ocean. It uses its big, strong claws to tear the flesh of dead animals. It also digs out the flesh of broken coconuts on the beach. Coconut crabs are well adapted to life on land; their gills have been replaced by lungs, and if they stay in the water too long, they will drown. Like other land animals, the crabs drink freshwater from inland springs and streams. They also climb trees to feed on juicy berries.

Horse

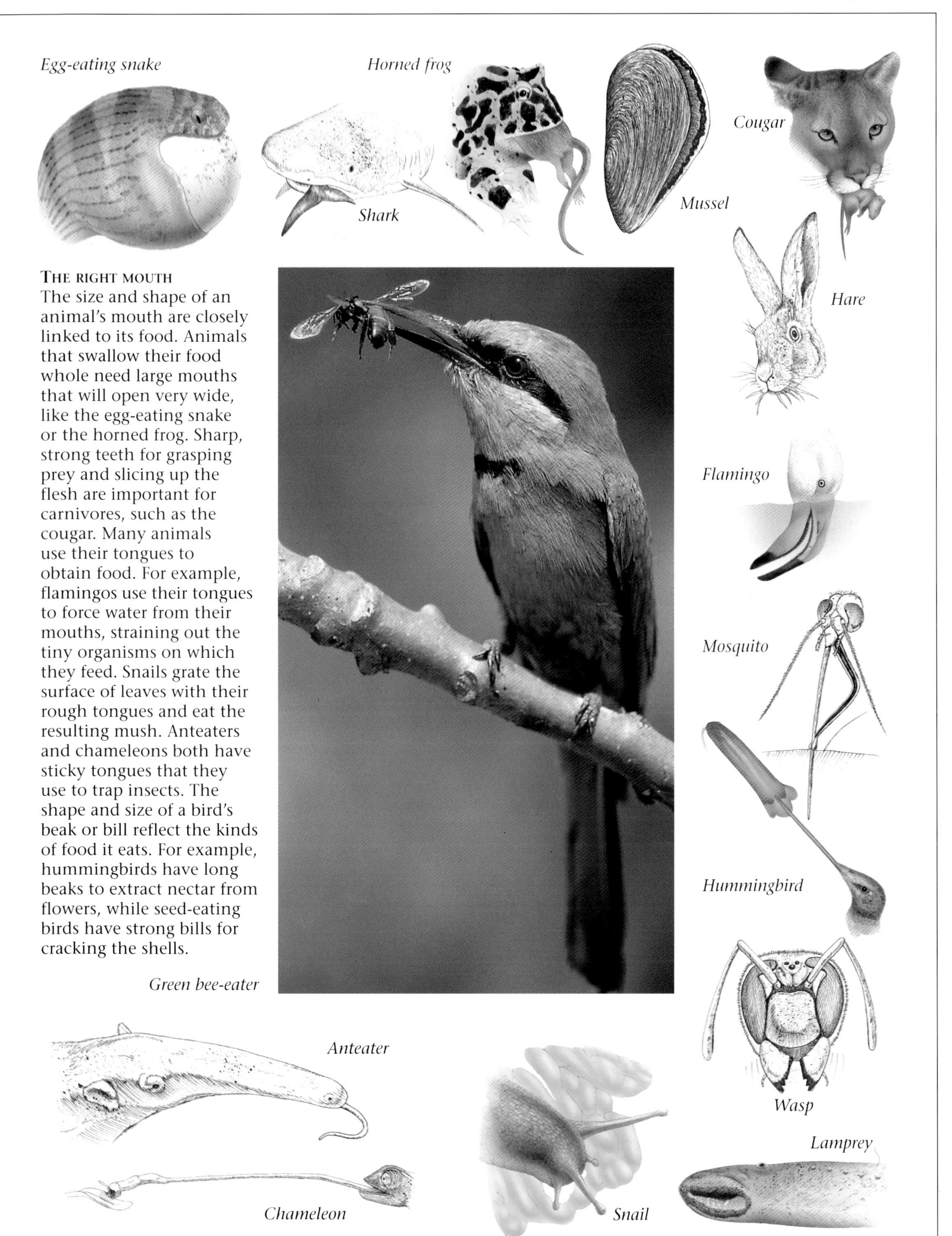

The right mouth
The size and shape of an animal's mouth are closely linked to its food. Animals that swallow their food whole need large mouths that will open very wide, like the egg-eating snake or the horned frog. Sharp, strong teeth for grasping prey and slicing up the flesh are important for carnivores, such as the cougar. Many animals use their tongues to obtain food. For example, flamingos use their tongues to force water from their mouths, straining out the tiny organisms on which they feed. Snails grate the surface of leaves with their rough tongues and eat the resulting mush. Anteaters and chameleons both have sticky tongues that they use to trap insects. The shape and size of a bird's beak or bill reflect the kinds of food it eats. For example, hummingbirds have long beaks to extract nectar from flowers, while seed-eating birds have strong bills for cracking the shells.

Green bee-eater

Plant-Eaters

Green plants get their energy from sunlight through a wonderful process called **photosynthesis**. They use this energy to make their own food. Animals cannot do this. They have to get their energy by eating plants or by eating animals that have already eaten plants.

Most of the world's animals are plant-eaters. Some eat a wide range of plants, but most stick to just a few kinds. They may eat only certain parts of plants, for which their digestive systems are well adapted. Leaves and stems consist mainly of a tough material called **cellulose**. This is not easy to digest, but **microscopic** bacteria and **protozoans** live in the stomachs of many plant-eating mammals and help them to digest it.

THE GIRAFFE'S JUNGLE COUSIN
The okapi was only discovered in 1901. The closest and only living relative of the giraffe, this shy animal lives in the dense tropical forests of central Africa. It feeds on leaves that it plucks from the trees with its long tongue. Its tongue can measure up to 20 inches (50 cm) in length—long enough to reach up and clean its own eyes! Okapis are picky eaters. They spend time searching for certain types of leaves, tender buds and shoots, ferns, fruits, and mushrooms. When they find vegetable gardens planted by human forest-dwellers, they particularly like to eat sweet potatoes.

FRESHWATER VEGETARIANS
The three species of manatees, along with their relative, the dugong, are aquatic mammals that feed almost exclusively on plants. They eat sea grasses and seaweeds. Manatees have a special **adaptation**: because the plants contain an abrasive substance called silica, which wears teeth down quickly, the animals shed their worn teeth throughout their lives, and new ones grow in their place. Manatees live in the shallow waters and rivers around the Atlantic coasts of West Africa and tropical America.

■ **More about okapis**
With its long legs and body sloping downward from the shoulders, the okapi looks like a giraffe with a short neck. Its dark brown coat and striped legs and hindquarters help it to blend into the forest. Okapis stand about 5.2 feet (1.6 m) high at the shoulders and can weigh up to 507 pounds (230 kg). The males have two small horns and, similar to giraffes, they fight for females by exchanging blows with their necks. The males stake out a **territory** of a few acres. Males and females meet only at mating time. The females bring their calves up alone. The calf stays with its mother for about 12 to 18 months.

Tree hyrax
Although hyraxes look like large **rodents**, elephants are their nearest living relatives. Tree hyraxes live in the forests of Africa. Their padded feet with clawlike nails help them run up smooth tree trunks. Sticky secretions from glands on their feet give them even better grip. Hyraxes come out at night to munch on leaves, fruit, ferns, and, sometimes, insects and birds' eggs. These noisy mammals usually live in pairs.

Eucalyptus for dinner again!
The koala is an Australian **marsupial** that spends its whole life in the eucalyptus trees on which it feeds. Eucalyptus leaves contain substances that are toxic (poisonous) to most animals. The koala has solved this problem by having a very long **cecum** and by avoiding the most toxic leaves. Eucalyptus leaves are not very nourishing, and koalas have to eat large amounts of them to obtain the energy they need to survive. They eat about 3 pounds (1.3 kg) of leaves every day. Before the Australian government introduced protection laws for koalas, they were hunted extensively for their fur. Recently, disease and human encroachment on their territories have decreased numbers of koalas from several million to just a few hundred thousand.

Flowers and Fruits

Some animals have discovered that the most nourishing parts of plants are the flowers and fruits. Flowers produce dustlike **pollen** that triggers seed production when it reaches the female parts of another flower of the same kind. The pollen contains nourishing proteins, and many insects and other animals find it good to eat. Many flowers attract insects by producing extra-large amounts of pollen. The insects eat a lot of it, but also carry some to other flowers, where it starts the process of seed formation. So the arrangement is useful to the plants as well as to the insects. A seed is a tiny plant surrounded by a store of food that nourishes it in the early stages of growth. Animals know that there is good food in seeds, and many species, especially birds, have become specialized seed-eaters. Seeds are often protected by tough coats, but the seed-eating birds have evolved strong beaks that can open the seeds very easily. Many seeds develop inside tasty fruits that are designed to attract birds and other animals. The animals carry the fruits away to eat them and, because they scatter the seeds in different places, they help the plants to spread.

FRUIT-EATING BATS
Bats can be divided into two broad groups: fruit-eaters and insect-eaters. Fruit bats are generally larger, and most species live in the **Tropics,** where fruit is plentiful throughout the year. Fruit bats help trees and plants reproduce. The seeds pass through the fruit bat's digestive system and fall on the ground, where they sprout and grow into new plants. The biggest fruit bats are called flying foxes.

A SUCCESS STORY
Rats are the most widespread animals on our planet. Hitching rides on ships and other means of transportation, they have followed people to the ends of the Earth, settling on even the remotest islands. Rats are successful because they eat almost anything, even if they prefer seeds and fruit. Brown rat populations are made up of clans led by a dominant male with a **harem** of females, plus subordinate (lower-ranking) males. When there are only a few rats, they defend their territory fiercely, chasing intruders away. When numbers increase, outsiders are tolerated.

The mockingbird's winter diet

Mockingbirds live in many parts of North America. During the spring and summer, when their young need large quantities of protein for growth, mockingbirds feed almost exclusively on insects. In the winter, when insects are less plentiful, they feed on seeds and berries left on trees and bushes. Although these foods contain little protein, they contain lots of energy-rich sugars, which help the birds survive through the cold winter months.

Powdering its nose

South African mice are among the few nonflying animals that **pollinate** plants. They feed on the nectar of the protea flower, a plant common in South Africa. Some species of protea produce flowers that open close to the ground and smell like spoiled butter. The smell becomes stronger at night when the mice are out and about. Attracted by this odor, some species of mice crawl over the flowers and drink the sugary nectar that they produce. While drinking, the mice get pollen on their noses. When they pay a visit to another protea, they pollinate the plant.

More about protea flowers and rodents

The flat-topped mountain, called Table Mountain, overlooking Cape Town in South Africa, is home to an exceptional number of plant and animal species. Table Mountain has its own cloud cover (known as "the Tablecloth"), which is mainly responsible for its lush vegetation. Several species of protea grow on Table Mountain; one of the most beautiful, the *Protea nana*, has brilliant red blooms. Scientists, however, were less interested in the beauty of many blooms than in some species' habit of growing insignificant-looking flowers very close to the ground. The mystery was solved in 1978 when a biologist noticed a host of tiny paw prints near the low-growing flowers. This, and the many tunnels around the roots of the proteas, pointed to the fact that tiny rodents were feeding on the pollen produced by the flowers. This is another example of the way plant and animal species cooperate to the advantage of both. The flowers rely on the mice to move pollen from plant to plant, thus ensuring the continuation of their species, while the mice add a touch of something sweet to their diets.

Nectar-Feeders

Nectar is one of the richest and most concentrated sources of food in nature. It consists almost entirely of energy-rich sugar dissolved in water. Most flowers produce it in small amounts throughout the day. At the same time, they produce strong scents that attract insects. Nectar is often produced deep down in tubular flowers, and the insects have evolved highly specialized organs to reach it. Butterflies and moths, for example, have long, slender tongues similar to drinking straws that can reach the bottom of the deepest flower. While probing for nectar, the insects unknowingly pick up pollen and carry it to the next flower, where it triggers seed production. Some hummingbirds and other birds also feed on nectar and help pollinate the flowers.

BUSY BEES
Bees were among the first animals to be domesticated by humans who realized the value of honey thousands of years ago. Bees feed on pollen and nectar from flowers, some of which is carried to the hive by female worker bees and turned into honey. When a worker bee finds a rich source of nectar, it zips back to the hive and tells the other bees where it is by means of a series of rhythmic movements. Scientists call this the "dance of the bees."

BATTY ABOUT NECTAR
Large tropical bats, called flying foxes, live in hot areas throughout the world. They eat fruit from a wide variety of plants, helping to spread the plants far and wide by scattering the seeds. One group of flying foxes are specialized nectar-feeders. They use their slender, brushlike tongues to reach deep inside the nectar-producing flowers. By unknowingly carrying pollen from plant to plant, they also help pollinate flowers. If these bats were to become extinct, many tropical plants would also go extinct.

KEEPING PACE
Like some hummingbirds, butterflies and moths help pollinate flowers. Thus, flowers benefit from their presence. The evolution of many species of plants has gone hand in hand with that of the butterfly. Changes in the structure of a flower were immediately followed by changes in the structure of the insects. An orchid in Madagascar, for example, has its nectar hidden right at the end of a 10-inch (25-cm) tube, and only one kind of moth is able to reach far enough inside to drink it.

Many butterflies and moths have a long, hollow tongue used to extract flower nectar.

More about lorikeets
Lorikeets form a distinct group within the parrot family. They differ from other parrots in that they mainly feed on nectar and pollen from flowering trees and shrubs. They are also sometimes referred to as honey parrots. Lorikeets dwell in Indonesia, Australia, New Guinea, and the Pacific Islands. Most species are brightly colored, and most live in groups. The male lorikeets frequently perform a series of elaborate dances to chase away rivals or to impress females during courtship.

Specially adapted tongues
Compared with other birds, parrots have large, fleshy tongues, which they use to move food around in their bills. Some species of nectar-feeding parrots have other adaptations to make them more suited to their diets. Lorikeets have a tuft of threadlike projections on the tips of their tongues. When they are resting or eating fruits or seeds, they keep the tufts covered in a sheath. However, when they find a flower that has nectar, the projections are extended into the flower to mop it up.

Suspended animation
Certain hummingbirds are the most specialized nectar-feeders among birds. Their beaks are very long and curved so that they can reach inside trumpet-shaped flowers. Their special way of flying allows them to hover in front of a flower as they feed. The hummingbird's tongue is specially designed for drinking nectar. Although long and strawlike, it is not used to suck the nectar but acts more like a sponge absorbing the nectar. The tongue is controlled by very long muscles stretching from the skull to the nostrils. Some hummingbirds are territorial and fiercely defend the nectar-rich plants in their area from other birds.

Boring for Food

Some small animals do not simply eat plants; they live most of their lives munching away right inside the plants. Animals that live like this, receiving both food and shelter from the plants, are sometimes called **parasites**. The **larvae** of many tiny insects, known as leaf miners, can live inside a single leaf. Brown patches or twisting tunnels show where they have chewed their way through the soft, nutritious tissue between the upper and lower surfaces. They bite their way out and escape when they are fully grown. The grubs (larvae) of many beetles and flies and the caterpillars of several moths live inside plant stems. Some even manage to live in dead tree trunks, but wood is not a very nutritious substance, and insects feeding on it often take several years to grow. Some plants have evolved chemical defenses against grazing and tunneling animals. Many of them contain **tannins** and other bitter-tasting substances, but not all plant-eaters are put off by them. Some actually use the poisons by storing them in their bodies and using them to deter their own enemies.

Bark beetles in the forests
Bark beetles live under the bark of trees, in the nutritious tissue called **phloem**. Males and females meet on the trunks and then bore through the bark. The female digs a little tunnel between the bark and the wood and lays her eggs in it. When the eggs hatch, the grubs make their own tunnels in the phloem. They eat it as they go and gradually spread out from the original tunnel. If the bark falls off later, you can see the fascinating tunnel patterns. Each kind of beetle makes a different pattern. When they are fully grown, the grubs turn into new beetles, which chew their way out of the bark to go in search of mates and new trees. Too many bark beetles can cause serious damage to the trees.

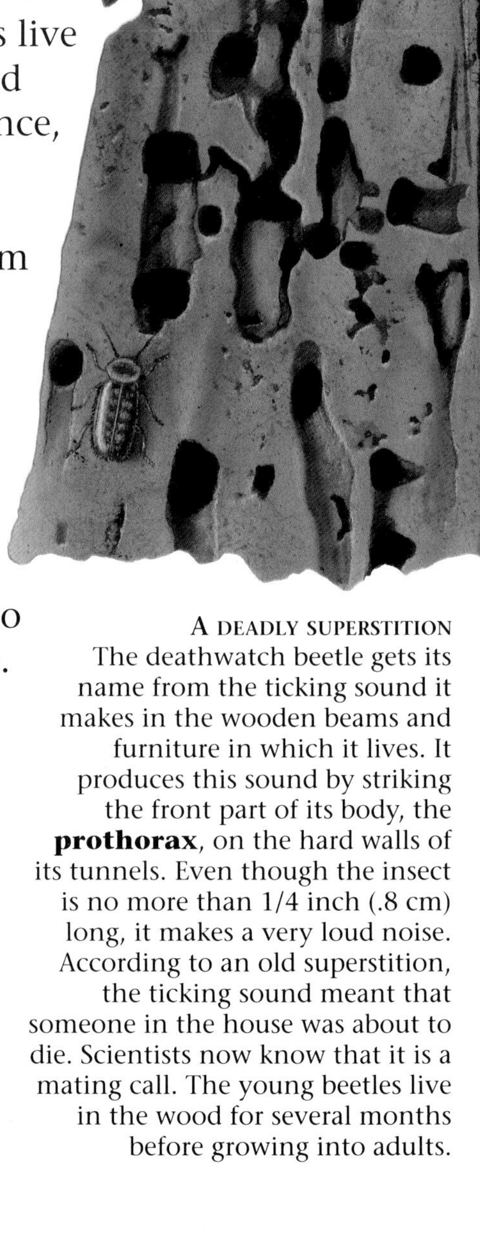

A deadly superstition
The deathwatch beetle gets its name from the ticking sound it makes in the wooden beams and furniture in which it lives. It produces this sound by striking the front part of its body, the **prothorax**, on the hard walls of its tunnels. Even though the insect is no more than 1/4 inch (.8 cm) long, it makes a very loud noise. According to an old superstition, the ticking sound meant that someone in the house was about to die. Scientists now know that it is a mating call. The young beetles live in the wood for several months before growing into adults.

Poplar borers
The poplar borer is a beetle with grubs that make long tunnels in the branches of poplar trees. The tunnels weaken the smaller branches, making them especially prone to wind damage. To defend themselves, the trees produce growths similar to galls (see below). In severe infestations all the smaller twigs and branches up to 1/2 inch (1.5 cm) thick develop one or more swellings. The insect larvae inside the swellings take one or two years to become adult, growing to 1 inch (2.5 cm) in length and 1/4 inch (.8 cm) in diameter. Other wood-boring beetles cause damage by helping fungi to enter. The fungi cause canker, a disease that can quickly kill the trees.

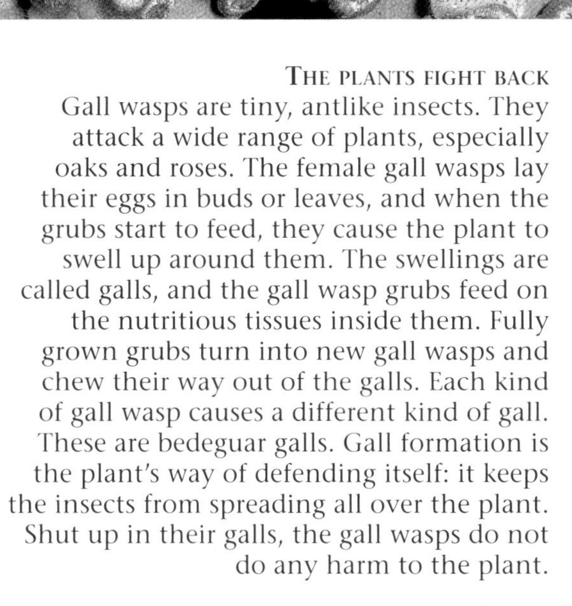

The plants fight back
Gall wasps are tiny, antlike insects. They attack a wide range of plants, especially oaks and roses. The female gall wasps lay their eggs in buds or leaves, and when the grubs start to feed, they cause the plant to swell up around them. The swellings are called galls, and the gall wasp grubs feed on the nutritious tissues inside them. Fully grown grubs turn into new gall wasps and chew their way out of the galls. Each kind of gall wasp causes a different kind of gall. These are bedeguar galls. Gall formation is the plant's way of defending itself: it keeps the insects from spreading all over the plant. Shut up in their galls, the gall wasps do not do any harm to the plant.

Great blue hunter
The blue shark, found in oceans throughout the world, is one of the fiercest predators. It can grow up to 16 feet (5 m) long and weigh more than one and a half tons. Like all sharks, it uses its highly developed senses to locate prey. It can detect the slightest odors in the water. Blue sharks sometimes strike seals with enough force to knock them senseless. The sharks then use their razor-sharp teeth to slice up their victims.

A Diet of Meat

Meat-eaters, or carnivores, form the last links in the food chains or in every ecosystem. Food chains always start with plants, because only plants can make their own food. The next link in the chain is a plant-eater, getting its energy from the plants that it eats. Then come one or more meat-eaters that get their energy by eating other animals. The final link is often a large and powerful animal, such as a shark, a wolf, or a big cat. These top predators can never be very common, because there would not be enough food and energy to go around. Energy is lost at every stage of the food chain, because the animals use it up all the time. When an antelope is caught and eaten by a leopard, it contains only a small fraction of the energy that it has taken in during its lifetime. Food or energy chains are sometimes called pyramids, because there are lots of plants at the base, smaller numbers of herbivores farther up, and even smaller numbers of predators at the top.

Getting a grip on things
The giant anaconda is a snake that grows to about 16 feet (5 m) long. Reports exist of specimens measuring as much as 40 feet (12 m), but there are no official records of these. Anacondas live along tropical rivers in South America. They lie in wait along riverbanks for prey, which they kill by constriction (squeezing in the coils of their body until the prey can no longer breathe). Anacondas can kill creatures as large as pigs, tapirs, and caimans. They are excellent swimmers and will follow their prey into the water.

The starfish closes itself around its victim, in this case a mussel.

Its suckers start to pull the shell apart. The starfish's gastric juices gradually weaken the mussel's resistance.

The mussel opens slowly. The starfish pushes the rest of its stomach into the shell and digests the mussel.

A death star

Starfish are spiny-skinned animals belonging to the group called **echinoderms**. They are carnivores, feeding on a wide range of other animals. They also eat **carrion**. Powerful suckers under their arms help them move and feed. The suckers can open the shells of **mollusks**, such as cockles and mussels. Digestive juices released by the starfish attack the muscles holding the shells shut, so the shells gradually open. The starfish then pushes its whole stomach into the mollusk to digest the flesh.

More about starfish

There are more than 1,800 species of starfish. They live in seas and oceans throughout the world. The largest number of species lives in the North Pacific, from Puget Sound to the Aleutian Islands. For example, Vancouver Island, off the coast of Canada, has about 70 species living in its waters that are not found elsewhere. Most starfish have five arms, although there are some with more than forty. Starfish vary in size from about 2 inches (5 cm) up to 3 feet (1 m).

Lone stalker

The leopard is the most widely distributed of the big cats. It is found over most of sub-Saharan Africa, parts of North Africa and the Middle East, much of southern Asia and the Far East, as well as the snowscapes of Siberia. Leopards hunt by stalking and ambushing a wide range of prey, from small rodents to large herbivores, including antelope and deer. Hunting usually takes place at dusk or at night, with the leopard stalking its victim for a long time before closing in for the kill. Like most of the big cats, the leopard hunts alone. Females raise their young by themselves, and hunting lessons are a basic part of cubhood.

Patient Predators

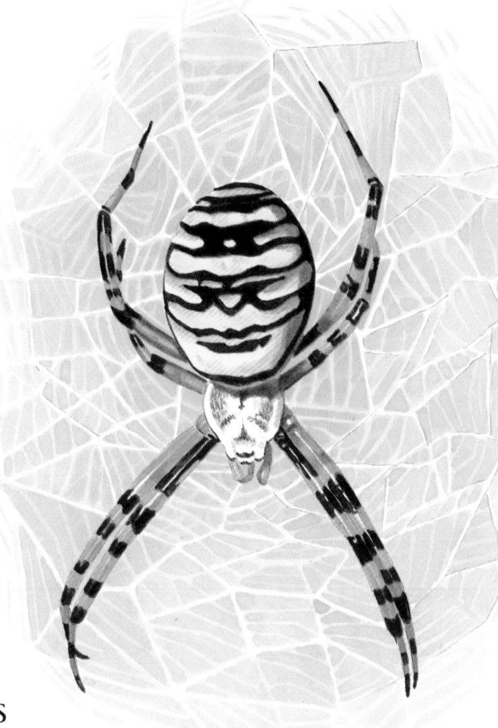

Not all meat-eaters and not even all predators are exceptionally strong or fast. They are only as big and strong as they need to be to catch their food. Many of them feed on small creatures and rely more on patience to find and capture them. Crab spiders often sit in flowers and wait for insects to arrive. The spiders are often the same color as the flowers, and the insects do not see them until it is too late: the crab spiders grab them and suck them dry. There are many other kinds of spiders, including the orb weavers whose silk, wheel-shaped webs are marvels of design and engineering skill. They can catch and hold insects much bigger than the spiders themselves, and the webs of some tropical spiders can even catch birds. The gladiator spiders are perhaps the most specialized of all predators. They spin sticky webs and throw them over passing insects.

Pouncing on prey
The North American bobcat is slightly smaller than its cousin, the lynx. Similar to the lynx, the bobcat hunts by suddenly leaping onto its unsuspecting victims. It eats birds and small mammals, such as snow hares. In the winter, when prey is scarce, it will sometimes eat carrion.

THE ART OF WAITING
The striped argiope spider spins a very large and extremely strong web in the grass or bushes. The females spin the largest webs. The spiders wait head-down in the center of their web. When an insect is trapped, they race out and wrap the victim in silk thread. The prey is killed by the spider's **venom**, which also helps to break down the food.

A COLORFUL AMBUSH
Horned frogs are among the most aggressive species of frogs. These amphibians are well camouflaged by their bold patterns as they sit perfectly still in the grass or swamp water. They eat almost anything that will fit into their large mouths, from worms and insects to mice, lizards, and small frogs. Because of the size of their mouths, which stretch almost to their legs, they are known as "the mouth with legs." The horned frog's bite is strong and painful because of its upper teeth. Horned frogs live in tropical America and Southeast Asia.

MORE ABOUT SPIDERS
Spiders are not insects; they belong to a group of animals called arachnids, which have four pairs of legs and two main body parts. They also have poisonous fangs. Spiders are found on every continent except Antarctica. Over 34,000 species of spiders have been discovered so far. They range in length from less than .04 inch (1 mm) to about 3.5 inches (9 cm). Spiders are carnivores and prey mostly on insects. Many spiders spin silk webs or threads to catch their prey, but many others, including wolf spiders and jumping spiders, chase and pounce on their victims. In many species the female is larger than the male and, unless he is careful to identify himself, the male will be mistaken for prey and eaten when he tries to mate with a female. The females lay their eggs in silk sacs. The young spiders do not look very different from their parents, except in size.

NICE, BUT FATAL!
Despite their appearance, ladybugs are efficient and eager predators. Most species feed on aphids (small insects that suck the juices from plants), which they find by methodically patrolling plants. They climb to the top of a shoot, and if they find no aphids, they go down again and climb the next branch they meet. In this way, they cover the whole plant. The seven-spotted ladybug is the most common species. During its brief lifetime as a larva and adult, it devours more than 6,000 aphids. The larva alone eat more than 2,000 aphids. The insect's bright coloring acts as a warning to other predators that know these tiny beetles have poisonous liquids inside.

Anything Goes

Most carnivores will eat any kind of meat that comes their way. This is especially true of the smaller meat-eating mammals. They feed on other mammals, birds, reptiles, amphibians, and insects. They also eat eggs and carrion. If they are really hungry, they even turn to plants. Because they take every opportunity to feed, these animals are called **opportunistic feeders**. They are more adaptable than the bigger and more specialized meat-eaters, and their digestive systems are less specialized. They can easily change their diet when food sources change. In the fall, for example, the opportunists often eat a lot of fallen fruit. The fox is one of the best known of the opportunistic feeders. It is so good at adapting to different environments that it is just as common in towns as it is in the countryside. Many flesh-eating birds are also opportunistic feeders.

THE AMBUSH AND STAB HUNTING TECHNIQUE
The gray heron is the largest and most widespread of the European herons. It nests from Great Britain to Greece, almost always close to water. It also lives in Africa and Asia. The gray heron normally stays close to marshy areas and does most of its hunting there, although it will venture into stubble fields or even farther afield if it has no young to feed. The gray heron eats a wide variety of animals, including frogs, moles, rats, and fish, which it captures with an ambush and stab technique. The heron waits on the banks of a lake, river, or swamp until an unsuspecting victim peeks out of a hiding place. Then, with the quick darting movement of its powerful neck, it stabs its victim.

WHEREVER THE FOOD IS
Most seagulls are highly intelligent. They take advantage of every opportunity that arises to get food. Some individuals within a species learn specialized hunting skills. For example, some gulls prey on eggs and young birds while others become specialist thieves, stealing fish from the beaks of other seabirds. Recently, seagulls have learned that where people live there will often be food. Garbage dumps and newly plowed fields have become important sources of food. Some seagulls have learned to live in cities, sometimes even nesting there, far away from their natural habitat.

Using its brain

The raven is one of the most intelligent birds. It feeds on anything it can find, from eggs and carrion to rodents, young birds, and even the **placentas** of mammals who have just given birth. Some scientists claim that ravens exchange information on the whereabouts of prey.

The mongoose picks up the egg...

Get cracking!

Mongooses are active predators. They feed on small mammals, birds, reptiles, eggs, and even fruit. This small opportunistic predator is a cunning creature: when it wants to eat an egg, it takes the egg in its front paws and tosses it backward between its legs, against a rock, to crack it open.

Hungry coyote

The coyote is one of the most widespread carnivores in North America. Very adaptable, it will make a meal of any piece of meat it can sink its teeth into, from carrion and rodents to young deer or elk. Working together as a pack, coyotes will also attack larger animals. When no meat is available, they will eat vegetables.

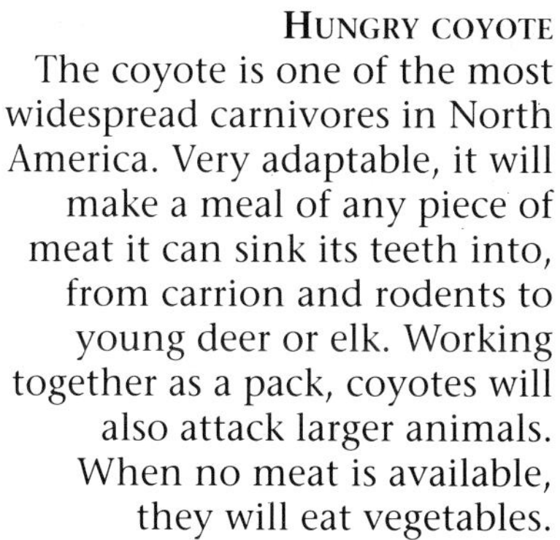

throws it backward between its hind legs...

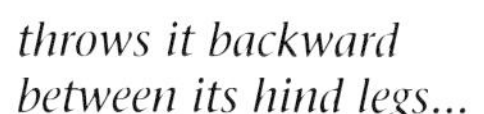

and dinner is served!

More about mongooses

Mongooses are small, carnivorous mammals with short legs, a pointed nose, small ears, and a long, furry tail. There are more than 75 species living throughout Africa, Asia, and southern Europe. Adult size varies considerably, ranging from 12 to 17 inches (32 to 44 cm) in the dwarf mongoose to 3 to 4 feet (1 to 1.2 m) in the small-toothed mongoose. Some species live alone or in pairs, while others gather together in large groups. Mongooses live in burrows and hunt during the day. Most species live on dry land, although one or two also spend some time in the water. They are bold hunters, and some species will even attack and kill poisonous snakes. They are not immune to snake venom and rely on speed and agility to avoid being bitten.

Insect-Eaters

Insects are the most abundant animals on Earth, and they are a good source of food for many other creatures. Because insects are all rather small, the insect-eaters themselves are usually fairly small. Many small birds are specialized insect-eaters. They have superb eyesight for spotting the insects, and most have sharp, tweezerlike beaks for picking them up. Woodpeckers have powerful beaks that they use to drill into tree trunks to find wood-boring insects. Swallows and nightjars catch insects in flight by scooping them up in their gaping beaks. Bats feed in much the same way. Larger insect-eating mammals specialize in the ants and termites that live in large colonies. Large numbers of insects can be caught at one time with very little effort. Anteaters have strong claws for ripping open the ants' nests, and long, sticky tongues that mop up the insects. Many of them also have long snouts.

A SCALY SUIT OF ARMOR
Pangolins, or scaly anteaters, are curious creatures that have bodies covered with hard, overlapping scales. When alarmed, they can roll into a ball that few predators can penetrate. Four species live in Africa, and three more live in southern Asia. They feed on the hordes of ants and termites that inhabit the forests and the grasslands. They use their sensitive noses to find the nests, and rip them open with their powerful claws. Their sticky tongues, which can extend up to 16 inches (40 cm) in front of the mouth, pick up the insects. Pangolins have no teeth, so the insects are ground up in the stomach.

THE STRANGEST MAMMAL
The aardvark is one of the strangest mammals. It lives in many parts of Africa and has no close relatives. The daily life and habits of the aardvark are also little known. It sleeps in a burrow by day and comes out at night in search of prey. It uses its strong claws to rip open ant and termite hills, and then pokes its long nose in. It mops up the insects with its long, sticky tongue. Like many other **insectivores**, its teeth are weak, and food is "chewed" in its muscular stomach. The aardvark swallows sand and pebbles while eating to help grind the food.

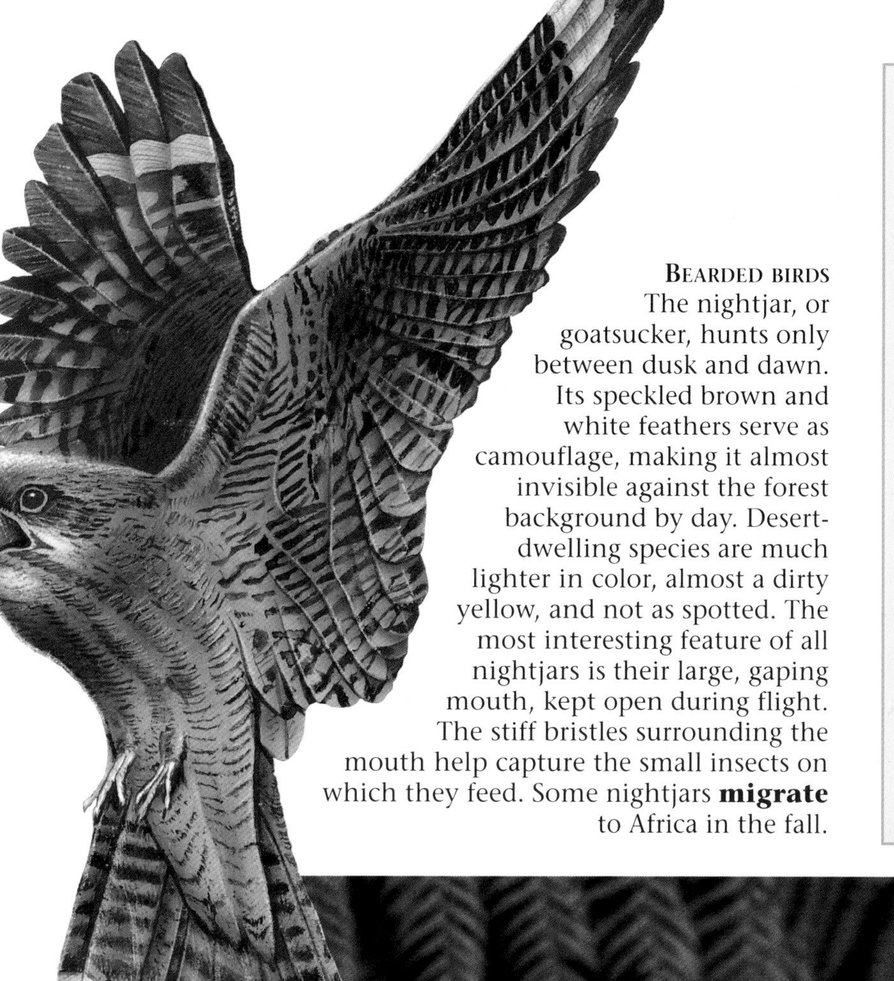

Bearded birds
The nightjar, or goatsucker, hunts only between dusk and dawn. Its speckled brown and white feathers serve as camouflage, making it almost invisible against the forest background by day. Desert-dwelling species are much lighter in color, almost a dirty yellow, and not as spotted. The most interesting feature of all nightjars is their large, gaping mouth, kept open during flight. The stiff bristles surrounding the mouth help capture the small insects on which they feed. Some nightjars **migrate** to Africa in the fall.

More about nightjars
There are over 80 species of nightjars spread throughout the world. They are sleek birds with large heads and tiny bills. There are few differences between the males and females, except in some tropical species, where the male is more brightly colored. Almost all nightjars are **nocturnal,** and some biologists think that, similar to bats, nightjars may use a form of **echolocation** (method of locating prey by bouncing sounds off them and other objects) in order to hunt more effectively in the dark.

Smells good!
The kiwi is found only in New Zealand. It cannot fly, and its tiny wings are hidden in its plumage (feathers). It moves through the forests on its long, muscular legs. The kiwi eats ants, earthworms, and other small **invertebrates** that it finds in the ground. Its beak is similar to the long snout of other insect-eaters. The bristles around the base serve as feelers. Its nostrils are at the top of the beak, and they help the bird find its food. Unlike most birds, it has a good sense of smell.

Animal Parasites

Some animals spend all or most of their lives on or in another creature, from which they take all their food. Animals of this kind are called parasites, and the animals that they feed on are called the hosts. Almost every animal is host to one or more parasitic species. Some parasites, including lice and fleas, stay on the outside of the host's body and suck blood from it. Others, such as tapeworms and roundworms, live inside their hosts. Some of them have become so specialized that they are hardly recognizable as animals at all. Many live in the host's digestive system. Some parasites settle in other parts of the body, such as the eyes or the blood vessels, where they can cause serious injury. Parasites rarely kill their hosts, but animals with lots of parasites often die because they are too weak to escape from predators.

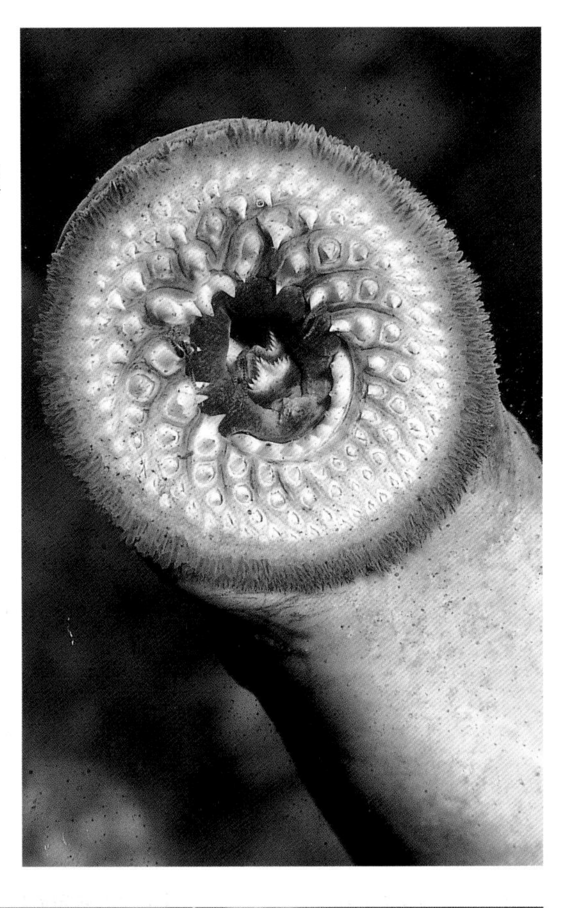

A jawless mouth
Sea lampreys and related freshwater species are fish parasites. Although they have long bodies like eels, they are more primitive **vertebrates** than fish. Their mouths have no jaws. They feed by attaching themselves to animals, usually fish, with mouths that are like suction cups. Long teeth surrounding the mouth scrape away the skin, and the mouth sucks blood and other liquids from the victim. When full, the lampreys detach themselves and rest on the bottom until hunger strikes again.

Riding on a fish
Many marine animals have become parasites. **Crustaceans** of the isopod group are among the most common parasites. They grip onto their fish or crustacean hosts, usually on their heads or backs. The *Anilocra physodes (*shown above) usually lives on fish that dwell on the ocean bottom or near the coasts. It is often found on fish that live among the seaweeds. This parasite is very common and is quite visible as it rides around on its host.

Dangerous guests
Tapeworms (shown right) are one of the most common parasites of human beings. These tapeworms spend part of their life in the stomach of a cow or a pig. Then they move into its muscles where they form hard lumps, called cysts. If a person eats this meat raw or undercooked, the cyst develops into an adult tapeworm, attaching itself to the stomach wall by means of little hooks on its head. It absorbs food through its skin and can live for a long time in the host's body, sometimes reaching 10 feet (3 m) in length. During its stay, it produces thousands of eggs that pass out of the host, ready to begin the cycle all over again.

A bloody meal
Leeches are related to earthworms, although they have adopted an entirely different lifestyle. They live as parasites by sucking blood from a variety of animals. Most leeches live in water, although there are some land-living species, particularly in tropical regions. Leeches are attracted to their hosts by the heat or chemicals they emit. They attach themselves to the host, cutting the skin slightly with their jaws; at the same time they inject a painkiller. Their saliva also contains a substance that keeps the blood from clotting.

CHARMING THIEVES
Raccoons are found practically everywhere on the American continent, from the east to the west coasts and from the mountains to the swamps of the south. They eat almost anything, from meat and plants to eggs. The raccoon is graceful and intelligent. Although its diet is mainly vegetarian, it will also climb a tree to raid a bird's nest, or, like the European badger, patiently hunt for earthworms. The raccoon's slender fingers are very good at handling food, and their fingertips are almost as sensitive as those of humans. No satisfactory explanation has so far been found for their habit of "washing" their food, although some zoologists think this is done only by animals in captivity.

Omnivores

Many herbivores occasionally eat flesh, and meat-eaters also eat vegetables from time to time. Foxes, for example, enjoy fallen fruit in the autumn. But there are many other animals that regularly eat both plant and animal foods. These animals are called **omnivores**, which means "eating everything." Cockroaches, ostriches, rats, pigs, and most bears are omnivores; so are most humans and chimpanzees. The teeth and digestive systems of omnivores are less specialized than those of carnivores and herbivores, because they have to digest both plant and animal matter. Omnivorous mammals have biting teeth at the front and chewing teeth at the back. However, they do not have big, stabbing teeth for killing prey, sharp-edged cheek teeth for slicing up meat, or big grinding cheek teeth like those of the grazing mammals.

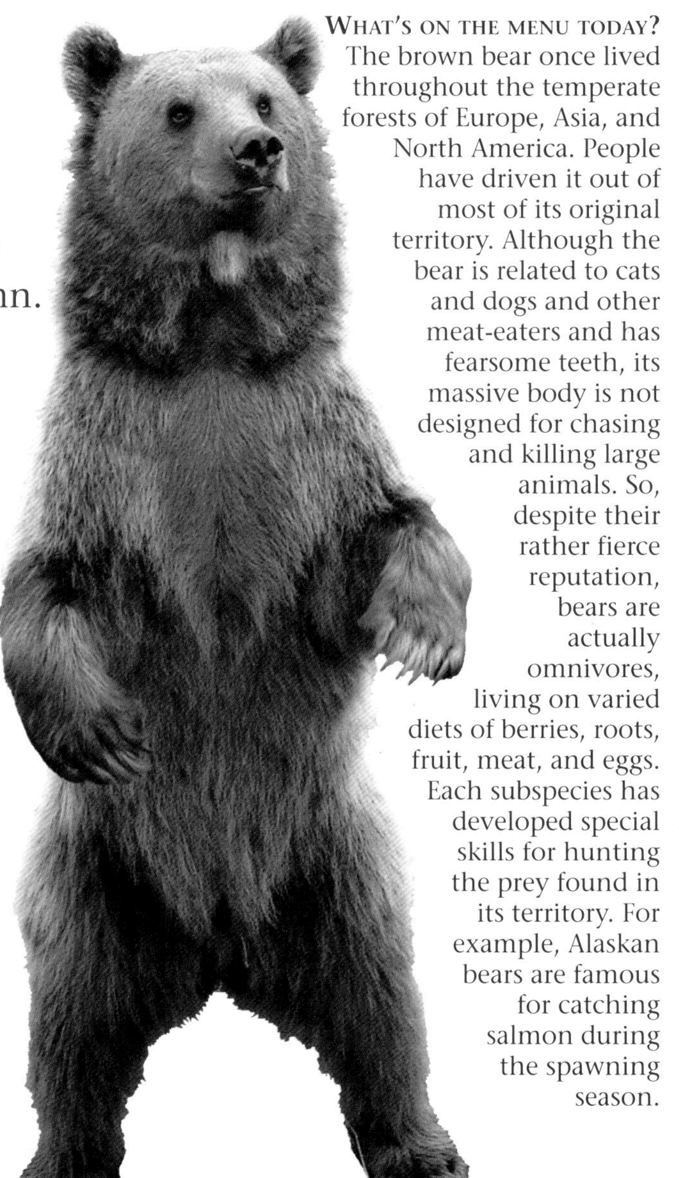

WHAT'S ON THE MENU TODAY?
The brown bear once lived throughout the temperate forests of Europe, Asia, and North America. People have driven it out of most of its original territory. Although the bear is related to cats and dogs and other meat-eaters and has fearsome teeth, its massive body is not designed for chasing and killing large animals. So, despite their rather fierce reputation, bears are actually omnivores, living on varied diets of berries, roots, fruit, meat, and eggs. Each subspecies has developed special skills for hunting the prey found in its territory. For example, Alaskan bears are famous for catching salmon during the spawning season.

A MOVABLE FEAST
The common crane, the only crane that nests in western Europe, is a powerful bird. During its long migrations from Africa to the woods of northern Europe, it eats everything, from potatoes in the fields of Spain to frogs in Polish swamps. Cranes always move in large groups and land only in areas where they are sure of finding food. This information seems to be handed down from generation to generation, and the cranes always follow the same migratory routes.

MIXED DIET
The blue-tongued skink is a reptile that lives in temperate Australian forests. It can grow up to 24 inches (60 cm) long. Like a few other reptiles living in colder climates, the skink does not lay eggs, but gives birth to live young. While most blue-tongued skinks are mainly vegetarian, some species (like the one shown here) eat a bit of everything, including worms, snails, insects, and other invertebrates. When threatened, the skink sticks out its blue tongue to frighten the predator. It also makes its body swell to make it seem larger, a technique that many animals use.

VERSATILE TEETH
Warthogs are related to European wild boars. They both have long, continually growing canine teeth used for finding food in their native habitats. The warthog is adaptable and lives in many wooded and grassy habitats in Africa. Continually on the move, it feeds on whatever plants it can find. In dry periods it uses its tusks and tough snout to dig up the ground in search of tubers and roots. Warthogs also eat carrion and small mammals. When eating, it kneels down in a curious position, although it is always on the alert for predators, such as leopards and lions. When threatened, it often takes refuge in the dens of aardvarks, from which only their long, dangerous tusks are visible.

Carrion-Eaters

Whenever an animal dies, there are lots of other animals lining up to feed on the dead flesh. They range from flies and beetles to vultures and hyenas. Carrion-eaters are often called **scavengers**, but this name is also given to animals that feed on other dead matter. Vultures and other carrion-eating birds find their food by sight, but other carrion-eaters pick up the strong smell that always surrounds dead animals. They need strong teeth or beaks to rip open the carcasses, although some of the smaller scavengers wait for the larger ones to finish eating before they move in to clear up the scraps. Scavengers also need strong digestive juices to destroy the germs that quickly invade carcasses and begin the processes of decay. It sounds gruesome, but the scavengers play a very important role in breaking down and recycling the dead matter. Without them, land would quickly be covered with dead bodies and other debris.

RAZOR-SHARP TEETH
Piranhas have earned a reputation as ferocious predators, although many species are quite small, and they often prefer to feed on dead or weak creatures that are easier to catch. Piranhas are common in the rivers of eastern and central South America. They have large, blunt heads with strong jaws and razor-sharp teeth that close in a scissorlike bite. Piranhas travel in groups and are drawn to the smell of blood. They usually prey on other fish and small mammals, particularly those that are weak or sick. They also eat the bodies of tiny herons and other river birds that drop out of nests overhanging the rivers. Hunting in a group, and while the mother is absent or distracted, they have been known to capture and devour baby caimans.

GRASSLAND GRAVEDIGGER
The marabou lives throughout most of Southern Africa. It feeds mainly on dead animals, but will also capture insects, fish, rats, and small birds. Marabous will sometimes help themselves to carrion other scavengers are feeding on and provoke ferocious fights among different species of scavengers. They often live near slaughterhouses. The long, naked pouch that hangs below the throat has nothing to do with food, but is used as a warning signal and during courtship.

Bone-crusher

With its mouth full of sharp teeth and hostile attitude, the Tasmanian devil looks quite savage. Yet, despite its appearance, it feeds on the carcasses of dead animals it finds during its nighttime wanderings. It has very powerful jaw muscles and teeth to crush the bones of carrion. Recently extinct on the Australian mainland, the Tasmanian devil is now found only in remote areas of Tasmania. It was hunted to near **extinction** by farmers who accused it of raiding their chicken coops and who also liked its meat, which they said tasted like lamb. Like most other Australian mammals, Tasmanian devils are marsupials.

More about king vultures

The king vulture is one of the most brightly colored members of the vulture family. It has a strong beak, which is perfect for ripping open dead animals and pulling off their skins and feathers. King vultures have bald heads. This means that they can stick their heads into a bloody carcass without the problem of getting their feathers caked with blood. They have rough tongues that they use to scrape meat from carrion bones. Their claws are not particularly sharp, since they are used for walking rather than for grasping prey. King vultures nest at ground level, often under a hanging rock. The incubation period lasts for about 5 weeks. When the young hatch, they are almost completely black. Their head colors gradually develop as the birds mature.

Scavenging in the forest

The king vulture is the least known of the South American vultures. It lives in tropical forests, flying low over the treetops as it searches for food. Whereas most vultures and other birds find their food by sight, the king vulture probably relies on its sense of smell, because it would be difficult to spot food under the thick canopy of branches. The bird's colorful head and showy wattles probably play a role in courtship.

Feeding on Debris

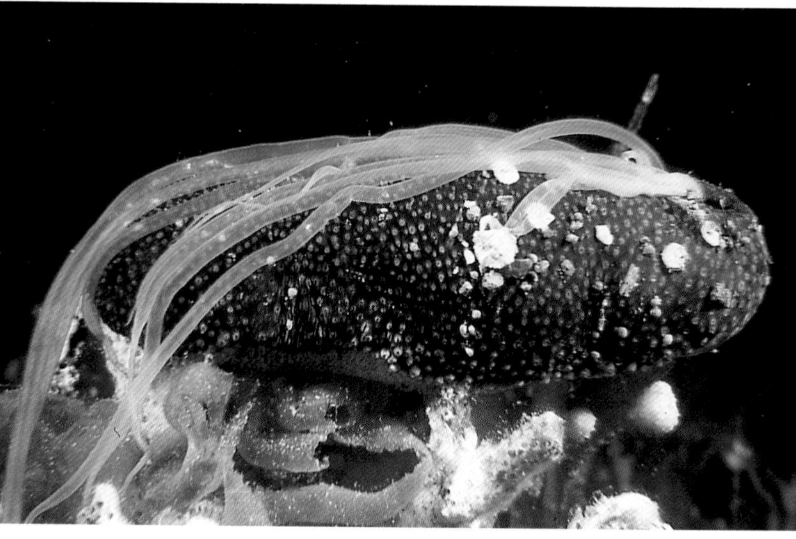

Dead plant material and animal dung on the ground or at the bottom of the ocean or a lake is eaten and recycled by a whole range of scavengers. It might seem easy to obtain food when all you eat is waste plant and animal matter, but there is not usually much digestible food in this detritus (debris). This is especially true of animal droppings, because most of the useful food has already been removed. The debris-feeders have to swallow large amounts of waste in order to get a small amount of food, or else they have to work hard to separate the good from the bad before swallowing it. Large animals certainly cannot get enough food and energy in this way, thus most of the debris-feeding animals are insects and other small creatures. Extracting the food from the debris is hard work. Water-dwelling debris-feeders pump large amounts of water through their bodies and filter food particles from it.

SEA CUCUMBERS
Sea cucumbers are echinoderms related to starfish and sea urchins. The sea cucumber's body has evolved into the shape of a long sausage. It lives on the sandy ocean bottom and between rocks, feeding on detritus. Sea cucumbers use the tentacles around their mouths to stir up the sand, which they then swallow. Any nourishing particles are digested. Many sea cucumbers stay buried in the seabed with only their tentacles showing. The feathery branches act like a net to catch the particles of food falling to the seabed.

A DIET OF DUNG
Scarabs are large dung beetles found in many of the warmer parts of the world. They feed mainly on herbivore dung, which is rich in vegetable matter. The beetles eat it themselves and use it to nurture their larvae. After digging a hole in the dung, the beetles lay a few eggs. When the larvae hatch, they find themselves in a safe "room" made of food in which they can feed and grow. In this way, dung beetles recycle undigested vegetable matter left by herbivores. They also return vital nutrients, such as nitrogen, to the soil where plants can use them to grow. Many scarabs bury the dung balls, after rolling them to suitable spots. They sometimes use carnivore dung as well.

Slow, but dangerous
Millipedes are common invertebrates living in soil and plant debris all over the world. Their body structure is fairly primitive, but they have large numbers of slender legs. They feed on plant and animal detritus. The structure of their mouth is quite simple; glandular secretions help them soften vegetable matter before eating it. In some tropical species, the mouth has been transformed into a kind of beak, used to pierce plant walls and suck out the sap. Although slow, millipedes are often dangerous. Many have **glands** that produce poisonous or irritating substances. Some tropical millipedes can spray these substances up to 12 inches (30 cm) far. But their bright colors warn other animals to keep away.

Tiny farmers
Earthworms are very important for the well-being of the soil. These small, segmented worms swallow soil as they tunnel and extract the nutritious particles. Undigested soil is passed out of the body and deposited as worm casts on or near the soil's surface. This benefits the soil because its lower layers are brought to the surface, and the top layers of organic waste sink into the soil where they can be completely decomposed by bacteria. Earthworm tunnels are very long and sometimes go down as far as 3 feet (1 m), especially in areas that are very dry or that have very cold weather.

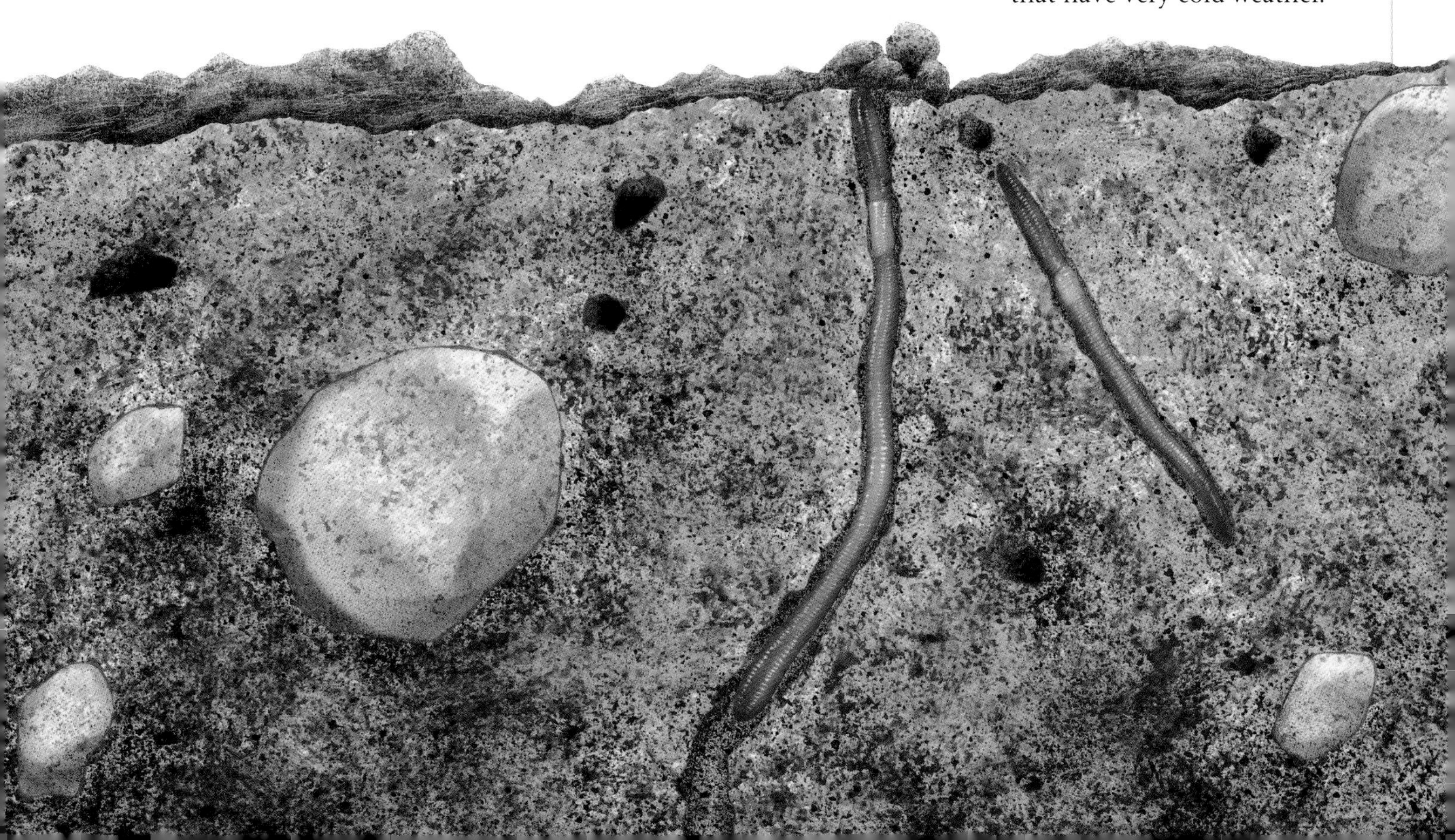

Filter-Feeders

Oceans, lakes, and ponds contain millions of tiny floating plants and animals, collectively known as **plankton**. They are sometimes so abundant that the water becomes cloudy. Lots of other animals feed by filtering the nutritious plankton from the water. A big whale swallows about 2.2 tons (2 tonnes) of it every day. Barnacles comb the little animals from the water with their feathery legs, while many other creatures suck in currents of water and strain the plankton from it.

KNEE-DEEP IN LUNCH
Although large, flamingos feed on tiny crustaceans in lakes and lagoons. Their filtering system resembles that of whales (though on a smaller scale). Water is sucked into the mouth, where a large muscular tongue pumps it through a kind of sieve lining the beak. The tiny crustaceans are held back and swallowed. In two varieties, the greater pink flamingo and the lesser flamingo, the filter plates differ in size so that they sieve out different prey. This potentially eliminates competition between the species.

OUR MARINE ANCESTORS
Sea squirts look like small, cylindrical bags with two openings. They draw water in through one opening, filter food particles out, and then discharge it through the other. Most sea squirts live permanently attached to the seabed. Their structure is so simple it is hard to believe that sea squirts are distant relatives of vertebrates.

UNDERWATER LACE
Sea fans are simple sea animals, although they look more like plants. They live in flexible tubes. Two feathery gills extend from the top of the tube. They capture tiny particles of food as they float by on the ocean currents.

*Humpback and other **baleen** whales strain plankton out of the water using the baleen plates that grow down from either side of the roof of their mouth. The whale opens its mouth wide, engulfing the plankton and large amounts of water. The water is then sieved through the baleen plates, separating the plankton, which is then swallowed.*

GENTLE GIANTS

The humpback whale is one of the largest animals on our planet, even though it feeds on tiny crustaceans that it finds in the cool Arctic and Antarctic oceans. Humpback whales concentrate in areas where there is an abundance of **krill**, which is their main source of food.

MICROSCOPIC SEA-DWELLERS

Under certain conditions some areas of the ocean are suddenly invaded by abundant growths of microscopic plants. This sets off a long and complicated food chain: within just a few days, there is an explosion of zooplankton, which eat the plants by filtering them through their gills or by other means. Zooplankton (right) are small animals that are carried along on ocean currents. The smaller ones are eaten by the larger ones, such as the shrimplike creature called krill (above). These are then eaten by fish and other larger animals.

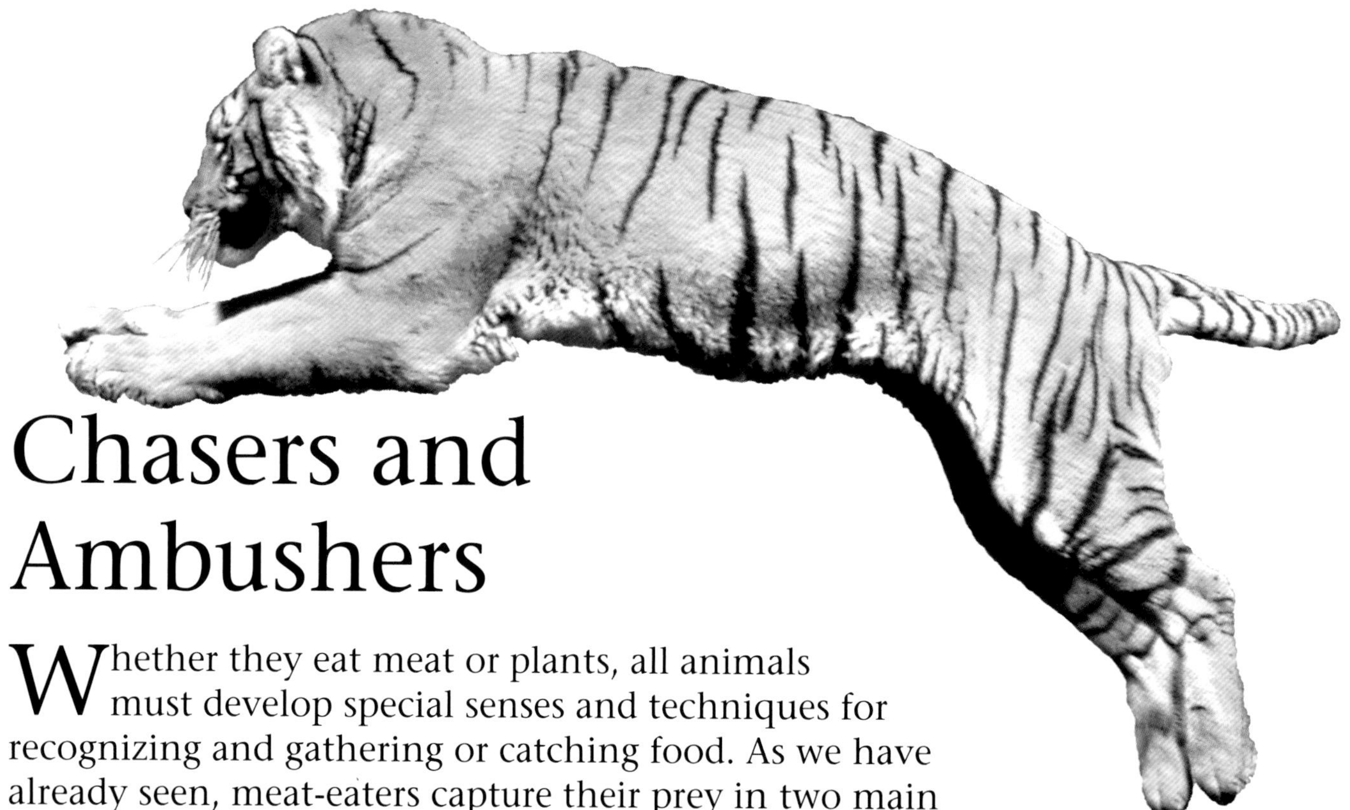

Chasers and Ambushers

Whether they eat meat or plants, all animals must develop special senses and techniques for recognizing and gathering or catching food. As we have already seen, meat-eaters capture their prey in two main ways—either by chasing it or by ambushing it. Chasers that chase their prey over long distances have muscles that can work for long periods without getting tired, but they cannot sprint at high speed. Cheetahs can sprint at high speed, but only for short distances. Mammals that ambush their prey are often heavier than the hunters. They leap onto their prey and kill it with the impact of their bodies as much as by biting it or using their claws. Ambushing spiders and insects also tend to be heavier than the chasers. The praying mantis, for example, has large, spiked front legs that it snaps around its victims. Many ambushers lure their victims within range with colorful fins or other parts of their bodies that act as bait.

Ruler of the forest
The tiger is the largest of the big cats. The Siberian tiger can weigh up to half a ton. Tigers live mainly in areas of long grass, where they can stalk and ambush all kinds of animals, from large antelope to small deer. Their striped coats, which blend in perfectly with the patterns of shade and light, and their stealthy movements allow them to approach their prey unseen. Similar to other highly evolved predators, tigers have developed different hunting methods according to their environment and their prey. For example, in Ranthambore Park in India, tigers have learned to attack sambar deer while they are in the water and cannot move fast. In Siberia, tigers chase deer into deep snow where they sink and cannot escape.

Anglerfish
There are many weird and wonderful creatures in the ocean depths! Many anglerfish, like the one shown left, have a luminous organ, called an esca, on their heads. The esca acts as bait to attract the fish they like to eat.

The voice of Africa
The territorial call of the osprey rings out over many lakes and waterways in Africa. Like its relatives in other parts of the world, this osprey hunts by diving down from a treetop (from where it has already spotted its prey) and grabbing the fish. Only its claws enter the water. Their feet are equipped with small nodules that help grip the fish as they fly away with it.

The jaws of death
At first sight, adult ant lions look a bit like dragonflies, although their antennae and wings are different. But the young stages of the two insects are very different. As soon as it hatches, a young ant lion digs a funnel-shaped hole in sandy soil. It then hides at the bottom and waits. It throws sand at ants and other small insects that stumble into the hole, and they slide down the sides, right into the ant lion's waiting jaws.

Weapons and Techniques

Capturing prey is not simply a question of having strong muscles or bursts of speed: chasers also need good senses and perfect coordination between the senses and the muscles. Ambushers need good coordination as well if they are to grab their victims at just the right time. The predators also have to kill their prey, and they have evolved just the right weapons to do it. Cats and dogs have large, stabbing teeth; birds of prey have sharp **talons**; and spiders and many snakes have venom. These features were not originally used for killing, but they have been modified during millions of years of evolution and are now deadly weapons. Insects and other invertebrate predators do not have to learn how to hunt or catch food, and spiders never need lessons in web-making. They are born with the necessary hunting skills and automatically know what to do. Behavior of this kind is called instinctive behavior. It is different among the hunting mammals, however. These youngsters have to learn by watching their parents, and they perfect their techniques with months or even years of practice in the family.

LOOKING BEFORE IT LEAPS
The kingfisher is the most colorful of European birds. Despite its bright coloring, it is not easy to spot when perched on a branch above a stream waiting for prey. The kingfisher feeds on small fish that it captures by diving into the water from above. Often, just before it dives, it hovers in the air beating its wings rapidly. When it hits the water, its eyes are covered by a protective membrane, and it can't see. So before it takes the plunge, it has to be absolutely sure of the position of its prey.

MORE ABOUT COBRAS
Cobras are venomous snakes that can expand their neck ribs to create a hood. They live in the warm areas of Asia, Africa, and Australia. The family includes the longest poisonous snake in the world—the king cobra of India—which can measure up to 18 feet (5.5 m) in length. Cobras mainly eat frogs, smaller snakes, rodents, and other vertebrates. Most species lay eggs that hatch into young.

Bird-eating spiders

Bird-eating spiders live in the warmer parts of America. Although many look alarming because of their size, their venom is not harmful to humans. Instead of spinning webs and lying in wait for prey, most species wander along the floors of tropical forests or climb trees actively hunting for birds and small mammals. Others dig burrows, line the walls with silk, and then wait for prey. If handled, the bird-eating spider defends itself by shedding its abdominal hair, which acts as an irritant and can cause severe swelling.

Giant of the airways

The golden eagle is one of the largest birds of prey in the Northern Hemisphere. It lives in the mountains and rugged areas of Europe, Asia, and North America. It can fly very fast when chasing prey and quite slowly when scouting an area. The eagle adopts different hunting strategies depending on the season, prey, and terrain. In the European Alps, for example, where the marmot is its preferred food, the eagle flies along at ground level in order to approach the rodents without being seen by their lookout. In other areas eagles feed on prey such as snow hares or partridges, which they pursue and capture at high speed. Golden eagles have been known to chase young chamois or mouflon goats over cliffs and to pick up their bodies at the bottom.

Raining venom

The spitting cobra lives in Africa. Like other members of the cobra family, its venom is highly poisonous. It is potent enough to kill large animals, including people.

The spitting cobra's teeth, which act as syringelike containers for the venom, have holes at the front, allowing the snake to squirt its poison over long distances. It uses its venom both to attack prey and to defend itself. When hunting, the venom quickly kills prey such as birds, mammals, or other snakes. For defense, the snake squirts it into the eyes of attackers up to several feet away, and this can cause blindness. As is true with all venomous snakes, the venom is a modified kind of saliva.

Stocking the Pantry

Food is not always readily available, and many animals have learned to store some for use in times of scarcity. Bears, dormice, and other animals that sleep for long periods in the winter usually eat well and put on weight in the fall. Their bodies gradually use the food reserves during the winter months. But storing extra fat in the body can slow animals down and make them easier to catch, so species that remain active in the winter usually store their food somewhere else. Squirrels and other animals that feed on seasonal nuts and seeds store them when they are in season and gradually eat them in the winter. The storage place has to be close to home but also has to be hidden from other animals that might steal the food. Many squirrels bury their stores in the ground or in hollow trees. Foxes and other predators even store meat for a while, especially in snowy places. Buried in the snow, the meat keeps well, just like in a freezer.

PICNIC IN THE TREES
Although leopards are among the most versatile of predators, they are not the strongest. African leopards have to contend with hyenas and lions trying to steal their food. To protect their catch from theft, leopards often drag it up into a tree where they can eat undisturbed. Using their powerful jaws, they can lift even quite heavy carcasses, including those of small zebras and gazelles. Keeping the meat in a tree also seems to discourage vultures that have trouble finding it in the branches. Asian leopards are less likely to eat in the treetops, because there are no lions or hyenas to threaten them.

BUTCHER BIRDS
The great gray shrike lives in Europe, North Africa, and North America. Although they are quite small birds (slightly smaller than a thrush), pairs defend extensive territories, sometimes as large as 100 acres (40 ha), in which they mate and hunt for food. Although great gray shrikes are songbirds, they share the hunting instinct and beak structure of birds of prey. They feed on small mammals, birds, and lizards. During the breeding season, they build special storerooms in thornbushes. The prey is hung on the thorns and taken down to feed the nestlings when required. This habit has earned the birds in this species the nickname of "butcher birds."

■ **More about woodpeckers**

There are almost 200 species of woodpeckers. They are distinctive birds, usually with small- to medium-sized, sturdy bodies and strong, sharp beaks. Most woodpeckers spend their whole lives in the trees. In the mating season many males use their beaks to tap communication signals on hollow trunks and branches. Most species feed on insects that they pick out from under the bark of trees with their pointed beaks. They help control numbers of bark and wood-boring insects, which might otherwise damage the trees. Many species dig nesting holes in tree trunks to create dwellings for rearing their young.

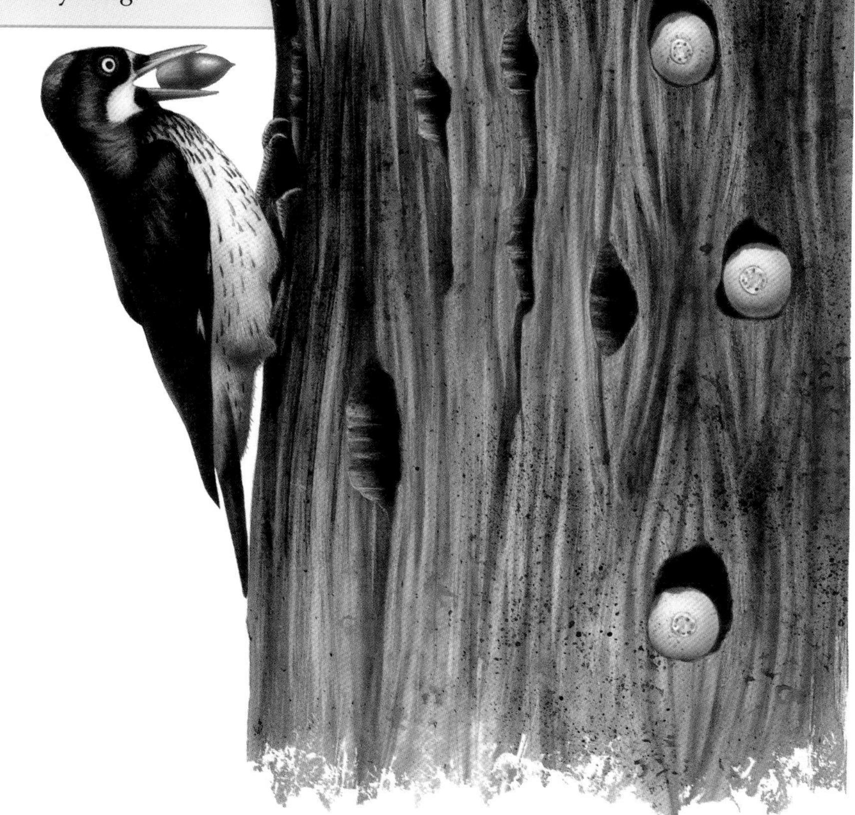

Designer storerooms

The American acorn woodpecker lives in groups of 3 to 12 individuals in **deciduous** forests from the southwest of the United States to Columbia, Canada. The birds in each group are usually part of the same family. They defend a territory of about 12.4 acres (5 ha), in which they mate and hunt for food. When they find acorns or other seeds, nuts, or even insects, that they can't eat immediately, they store them in dozens of small holes drilled in trees, telephone poles, or buildings. The acorns are well hidden, and only another woodpecker from the group can take them out. The storeroom is one of the most protected areas in their territory.

Underground storerooms

Moles dig very long underground tunnels for shelter and to search for their favorite foods, such as earthworms and other invertebrates. The tunnels themselves act as a kind of trap; worms breaking into a tunnel are easily caught and eaten by the moles. When it is necessary to store food, such as before fall and winter, the mole decapitates the earthworms and stores them in a special chamber. The largest store of earthworms ever found contained 1,200!

Watering Holes

The African savanna is hot and dry most of the time. Rain falls for a few weeks each year and fills the rivers, but then the long, dry season starts, and water becomes harder and harder to find. The lakes and rivers shrink, and only scattered watering holes are left. These draw animals from near and far like magnets. The first to arrive are those that need to drink most often—antelope and zebras. Giraffes and rhinoceroses are less affected by the drought at first, because they eat lots of juicy, green leaves. They usually come to the watering holes later in the dry season. The animals drop lots of dung, which attracts hordes of insects to the watering holes.

FIRST IN LINE
Antelope and wildebeests, among the most common savanna animals, travel toward the watering holes in long, dusty lines spread out across the savanna.

ELEPHANT-MADE HOLES
Sometimes the watering holes are made by elephants digging for underground water, which they can smell with their sensitive trunks. These pools soon fill up with mud and the elephant's waste. They are safer than most holes because there are no crocodiles or other predators lurking beneath the surface.

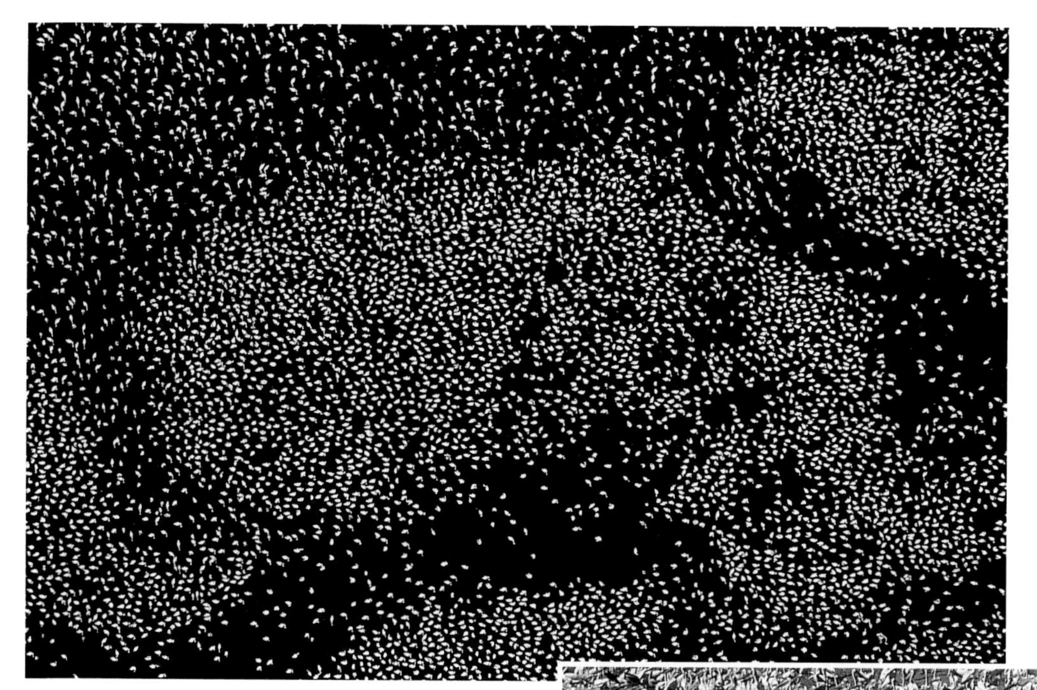

Pink lakes
Some African lakes have high levels of salt, soda, and magnesium deposits. Only a few species of crustaceans and the animals that feed on them, such as flamingos, can survive in these conditions. The pink pigments of the crustaceans cause the birds' feathers to become pink. Many of these lakes are ideal breeding grounds for flamingos, because they can raise their young protected from predators by the salty waters.

In search of fish
As the rivers and lakes dry up, the few fish left find themselves flapping around in just a few centimeters of water. At this point predators such as African fish eagles and pelicans arrive. They work the waters until not a single fish is left. When the water is completely dry, they fly off to another shrinking watering hole.

First come, first served
The order of arrival determines who drinks first, though some larger and stronger animals push their way up front. Then smaller herbivores, such as antelope and gazelles, have to wait for a free space to dip their noses in. The warthogs shown here (left) have had to make do with the murky, muddy water left by the zebras, although they don't seem to mind very much.

The Importance of Water

Food cannot be digested and used in the body without water. All other processes that go on inside the body also require water. It is used to carry away waste materials, and in mammals it also helps to control the body temperature through sweating. Meat-eaters get some water from the animal flesh that they eat, and plant sap provides some for herbivores, but the animals need to drink as well. Water is especially important for newborn animals. Baby mammals get water in the milk that they drink, and most baby birds get water in the juicy insects that their parents bring them. Sandgrouse travel many miles to find water for their nestlings and carry it back in their spongelike feathers. Very few animals can survive more than a few days without drinking. However, there are some beetles and scorpions that never drink, because they are able to conserve the water they get from their food.

MORE ABOUT STORKS
Storks are large wading birds with long legs and long beaks. They live in many parts of the world, although the largest number of species lives in Africa and Asia. Storks have long, broad wings and are excellent fliers. Unlike most birds, which feed by sight (first observing prey, then grabbing it with their beaks), many storks use another strategy, called feeding by touch. This involves keeping the bill open in the water and snapping it shut when something touches it. This method is effective when hunting at night or in muddy swamp waters, or where underwater vegetation is dense. Touch-feeding requires sensitive receptors on the outside of the bill and lightning-fast reflexes.

EGG WATERING
The shoebill is a large wading bird that lives in some of the marshlands of central Africa. Its large, multipurpose bill is its most striking feature. It is used as a shock absorber while hunting, when it launches its entire body weight against the reeds as it plunges after fish, small crocodiles, amphibians, and other animals. It also uses its beak to carry water to keep its eggs and hatchlings cool during the hot African summer. The eggs are watered four or five times a day and, when they hatch, so are the chicks.

Bathtime!
Cormorants are aquatic birds widespread on seacoasts and inland water systems throughout the world. The great cormorant of Europe and Asia is the largest of the species. It nests on rocky ridges or in treetops. Since the young cormorants are unable to get water themselves, their parents bring it to them. The parent birds carry the water in their throats and literally shower their young to cool them down. When the **hatchlings** need to drink, the parents open their beaks and let the water run into the young birds' throats.

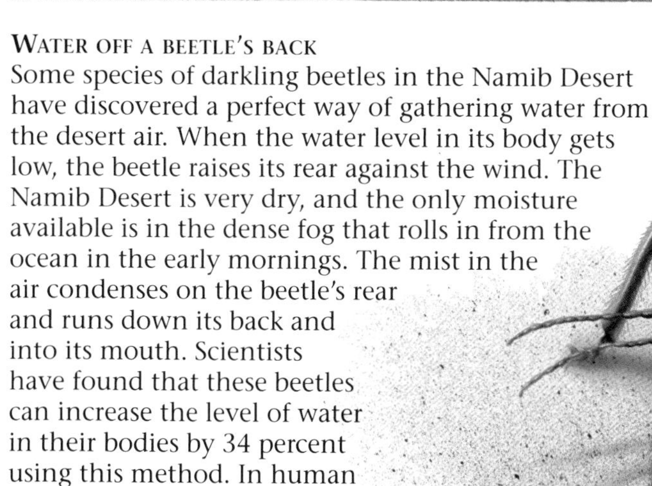

Dune cats
The sand cat, of the Sahara and the deserts of the Arabian Peninsula, never drinks water. It is able to survive because it gets the moisture it needs from the blood of the animals it hunts and kills. It feeds mainly on rodents that live at the bottom of the dunes where plant wastes are carried by the wind. The sand cat also eats small mammals up to the size of a hare, scorpions, reptiles, and insects. Together with the fennec fox and the Saharan leopard, it is one of the few carnivorous mammals that manages to survive in the barren desert.

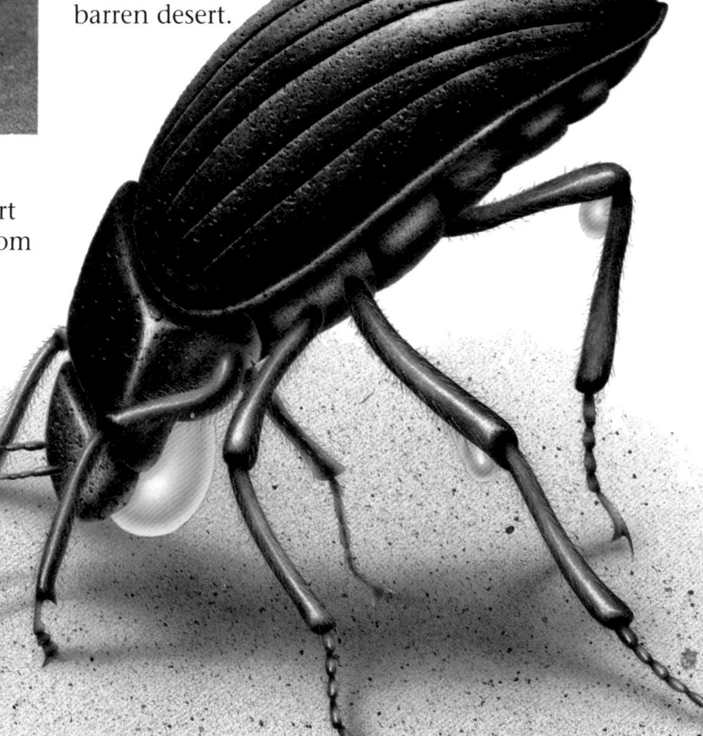

Water off a beetle's back
Some species of darkling beetles in the Namib Desert have discovered a perfect way of gathering water from the desert air. When the water level in its body gets low, the beetle raises its rear against the wind. The Namib Desert is very dry, and the only moisture available is in the dense fog that rolls in from the ocean in the early mornings. The mist in the air condenses on the beetle's rear and runs down its back and into its mouth. Scientists have found that these beetles can increase the level of water in their bodies by 34 percent using this method. In human terms, this would be the equivalent of a man weighing 154 pounds (70 kg) drinking 6.3 gallons (24 l) of water!

To the ends of the earth and back
The arctic tern breeds in the far north (sometimes within the Arctic Circle), where it feeds by snatching fish from the surface of the ocean in a series of rapid dives. When the mating season is over, the tern flies to the other side of the world to spend the northern winter on the pack ice just north of the Antarctic Circle. Here it uses the same hunting techniques as in the north, feeding on small fish that swim close to the surface. This small bird covers some 20,000 miles (32,000 km) each year. By passing from the Arctic summer to the Antarctic summer, it enjoys more hours of daylight than any other animal.

Migrating for Food

Migration is one of the most extraordinary events in nature. Huge groups of animals set off together on journeys that sometimes cover hundreds or even thousands of miles. The animals usually move when seasonal changes affect their food supply. For example, during the dry season the grass becomes too dry to eat in many parts of Africa, and the wildebeests and zebras walk hundreds of miles to new grazing lands. Later in the year, they move back again. But not all migrations are so dramatic: many mountain animals just move a thousand feet down the mountainsides in the fall and up again in the spring. When a population grows too big for the local food supply, many of the animals are forced to move to new areas, but in these instances they do not return. These movements, common for lemmings and locusts, are called emigrations.

Vanished migration
The springbok is one species that has paid dearly for human development. During the 19th century, the dry interior of Southern Africa was home to millions of springbok. In times of drought, when the grass had all but disappeared, the antelope would set off in search of food and water. They sometimes covered great distances, but when European settlers started to farm the land, things changed. The settlers kept the springbok from crossing their lands, and ended the great migratory movements. There are now far fewer springbok in Africa.

Flight of the butterflies
During the summer monarch butterfly larvae feed on the leaves of the milkweed plant in Canada and the United States. In the winter, the adult butterflies fly south. They fly north again in the spring, the females laying eggs along the way.

Migration on the steppes of central Asia
The saiga is a kind of antelope with a strange, swollen-looking snout. It has inhabited the steppes of central Asia for 20,000 years. Only the male saigas have horns, which they use to fight each other with during the mating season. Each male takes between 5 and 15 mates. In the winter, when the northern landscape is covered with snow and freezing winds blow, thousands of saigas go south to better grazing lands. In the spring they move north again. As they move, eagles try to catch young saigas that lag behind, and wolves wait to eat the placentas of the females after they have given birth.

More about saigas
During the last ice age, the saiga inhabited a vast area between Great Britain and eastern Siberia. It is now confined to the steppes of central Asia. Up until a few years ago, the saiga was faced with extinction, its numbers reduced to just a few thousand. A very successful campaign to protect the animals has allowed numbers to increase and the seasonal migrations to continue. Saigas are well adapted to their environment; their fine-boned legs and split hooves allow them to move at speed over the snow-covered steppes. The adult saiga stands about 30 inches (75 cm) at the shoulder. Its coat is short and brown in the summer and thick and light-colored in the winter.

Using Tools

Animals have always competed with each other for food. Evolution has produced many amazing adaptations designed to increase animals' ability to find and catch food. Various parts of their bodies have changed as each species has become specialized for catching and dealing with particular kinds of food. Changes in behavior have also been very important: some have enabled animals to move into different habitats and eat different kinds of food, while others have enabled them to obtain their food more efficiently. There are some astonishing hunting strategies in the animal world, but among the most surprising is the use of tools in finding and capturing food. It was once thought that human beings were the only animals that used tools, but scientists have now discovered that many animals use tools.

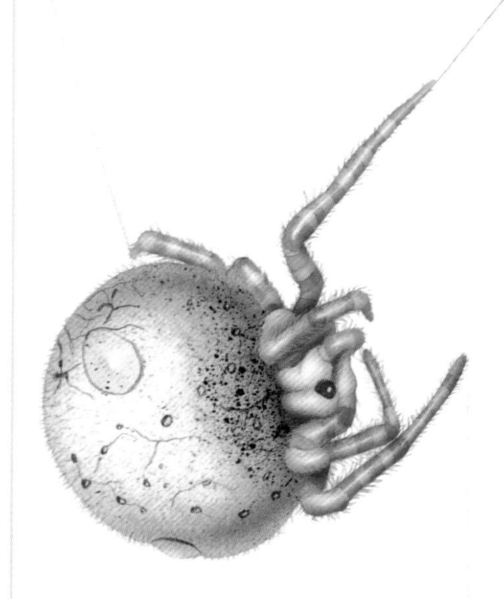

SILK FISHING LINE
Not all spiders build big, complicated webs to trap their prey. Some make do with just one sticky, silk thread. The bolas spiders are among the most skillful single-thread hunters. They live in many parts of the world, but the most common species comes from Australia. This predator positions itself on a plant stem and releases a length of silk with a drop of a gluelike substance on the end. When it detects a moth or some other likely victim, it whirls the thread vigorously around in an attempt to hit the prey. If it does, the gluey thread sticks to the prey's body, and the spider reels it in to eat. It seems that the spider also perfumes its "fishing rod" with the smell of flowers to attract prey.

DARWIN'S WOODPECKER FINCH
Darwin's woodpecker finch is one of several finch species descended from the few birds that colonized the remote Galapagos Islands thousands of years ago. The original birds adapted to their new environments by learning, among other things, to eat different foods from the ones they were used to on the mainland. Normally these birds feed on insects in rotting wood, but on the Galapagos one of them learned to use cactus spines, twigs, or small leaves to drive insects out of the wood.

PRECIOUS STONES
The sea otters that live along the Pacific coast of North America are famous for their use of a tool. When they pick up a large shellfish or sea urchin from the sea floor, they open the shell by banging it, usually while floating on their backs and holding a stone on their chests. Some otters even carry these stones around with them.

PRECISION FISHING
The archer fish lives in the mangroves of Southeast Asia. Mangrove swamps are a labyrinth of tree roots and debris washed up twice a day by the tide. They are home to thousands of animals, particularly insects that come to hide from predators at high tide. The archer fish has discovered a way of, quite literally, knocking insects out of the trees and into its mouth. The fish has a specially adapted mouth from which it squirts water at the insects with great precision. Its grey-green color and flat back allow it to blend in with its surroundings, making it hard for the insects to see from above.

BETRAYAL IN THE SHADE
The black heron lives in Africa, hunting for prey in swamps and ponds. It feeds on small fish, mammals, and aquatic reptiles, which it hooks out of the water with a quick flick of its neck and deadly beak. Its most curious behavior is not at the moment of capture but in preparing for the catch. The small heron moves slowly through the water with one or both wings raised, casting a shadow and preventing reflection. In this way it not only has a better view into the water but also creates a shady place where fish will instinctively feel safer. Attracted by the shade, the fish are easy prey for the heron.

Daily Food

The amount of food eaten in a day depends on a variety of factors, including the kind of food that is eaten. Some foods have more nutritional value than others, and animals need less of these foods. During the breeding season, female mammals need more food than usual, because they have to nourish the babies inside them. And when the babies are born, the mothers need extra food to produce their milk. Some birds change their diets from season to season, often eating buds in early spring, caterpillars in late spring and summer, and seeds in the fall. **Hibernating** mammals eat nothing during the winter, although they prepare themselves by eating well and gaining weight in the fall. Many insects eat only when they are young. Caterpillars, for example, spend almost all their time eating and growing, but when they turn into moths, many of them stop eating. Instead, they devote all their energy to mating and laying eggs. The evolutionary history of a species also affects its diet. An animal born with the digestive system of a carnivore cannot become an herbivore, because its digestive juices are designed for breaking down meat.

PICKY TICK (left)
The castor-bean tick is a hard-backed member of the arachnid family. It is an animal parasite that feeds on the blood of sheep. While feeding, the tick can take in an enormous volume of blood for such a small animal. It does this by expanding its abdomen to double its normal size, stretching from about .2 inches (4 mm) to .4 inches (10 mm).

THE BIGGEST APPETITE (top)
In the first 56 days of its life, the caterpillar of the polyphemus moth of North America eats food weighing 86,000 times its own birth weight!

A VEGETARIAN MEAT-EATER
The evolution of the giant panda is one of the most interesting and extraordinary among all mammals. The panda is related to the bear family, and like the bear, it has a carnivore's (or omnivore's) digestive system. The intestine is quite short, because flesh is easy to digest and carnivores do not need a long digestive system. However, for some unknown reason, the panda is an herbivore and feeds mainly on bamboo shoots. Because plant matter has a much lower protein content than meat, the panda has to eat a much larger amount of bamboo than a "real" herbivore would need to. The panda manages to extract just enough energy from the bamboo to stay alive. Pandas occasionally eat insects and rodents. Some zoologists claim that when food is scarce they will eat livestock, such as sheep, thus reverting back to their carnivorous nature.

THE COMMON GOOD
Lions are the only big cats that live together in groups in which the individual's behavior is organized and attuned with the needs of the others and with the pride as a whole. The pride, centered on several closely related females, stays together permanently, defending its territory from other groups. The group organization can be seen very clearly during hunting. Although some zoologists deny that there is any coordinated action, most people who have seen a group of lionesses hunting will agree that they are working together. A few lions will chase the prey (an antelope, zebra, wildebeest, or other animal) toward the others who lie in wait in the long grass. They leap out and kill the animal when it gets near them. Even large animals, such as giraffes or buffalo, can be caught this way.

Angry giants
The black rhinoceros is one of the largest animals of the African savanna. Its bulk, dangerous horn, and furious temper make it difficult for carnivores, even in groups, to capture. Using its long, flexible upper lip, it feeds on twigs and leaves of shrubs and trees, including the thorny branches of the acacia tree, which it breaks off with its horn. The black rhinoceros is able to recognize the nutritional value of certain plants and carefully searches them out. The white rhinoceros has a broad, flat snout designed for grazing. Because the two species eat different foods, they do not compete and live together in harmony.

On the shoulders of giants
Oxpeckers are small birds that spend most of their lives riding on the backs of large herbivores. They grip their hosts' skin with strong claws and use their beaks to pick off parasites and dead skin. Sometimes oxpeckers warn herbivores of danger by suddenly flying away. They live on giraffes, buffalo, rhinoceroses, zebras, and warthogs. Oxpeckers are an example of how two or more species can benefit one another by living in close proximity. This is called **symbiosis**.

Food for All

Animals do not live in isolation; they are all part of a vast web in which every kind of plant and animal plays its part. In the search for food, for example, animals can be helped or hindered by plants, by other animals, or by nonliving environmental factors, such as ice and snow. We have already seen how plants defend themselves against herbivores, and how herbivores protect themselves from carnivores: these are examples of how animals can be hindered in their search for food. A more subtle, but very important factor comes into play when two or more species compete for the same food. When this happens, either the weaker species die out or move away, or the different species "reach an agreement" and divide the food resources so that there is enough food for them all. Such a division of resources leads to specialized feeding habits, often with behavioral changes and sometimes physical changes as well.

■ **More about giraffes**

Giraffes are the tallest animals. Males can grow as high as 20 feet (6 m). Their long, flexible necks are useful for finding food that other animals can't reach; giraffes also use them to gain extra momentum and speed when running. By swinging them back and forth, they can reach speeds of up to 34 mph (55 kph). The giraffe's neck is often cited as the textbook example of evolutionary adaptation. Because a long neck gives them access to food out of reach of other animals, giraffes born with long necks have been more successful breeders and have gradually replaced ones with shorter necks.

Standing tall

The gerenuk is a slender, long-necked gazelle from East Africa. Although shorter than a giraffe's neck, the gerenuk's is long enough to give it an edge on other animals when feeding, for example, on the acacia tree. It reaches even higher by standing on its hind legs and pulling branches toward itself with its front legs. Unlike the giraffe, which needs to drink at least once a week, the gerenuk can go much longer without water. This means it can live in places that are too dry for other animals to survive.

Feeding at the top

Giraffes eat a wide variety of plants and, for this reason, have spread throughout many regions of Africa. The acacia tree is the giraffe's favorite food. Even the tree's sharp thorns are not enough to discourage the giraffe from nibbling at its tender topmost leaves. Giraffes have no competitors for their main source of food, because no other leaf-eating animal can reach it. They wrap their long tongues, which can measure up to 18 inches (45 cm), around the twigs, softening the tough leaves with saliva before plucking them. Giraffes also eat fruits, shoots, and seeds.

Pests and Guests

Many animals live in close contact with humans, taking advantage of the shelter we provide in our homes and the wide variety of foods that we eat. Most of these animals are omnivores, and they make use of every single scrap of food that we leave behind, as well as taking unused food from shelves and cupboards.

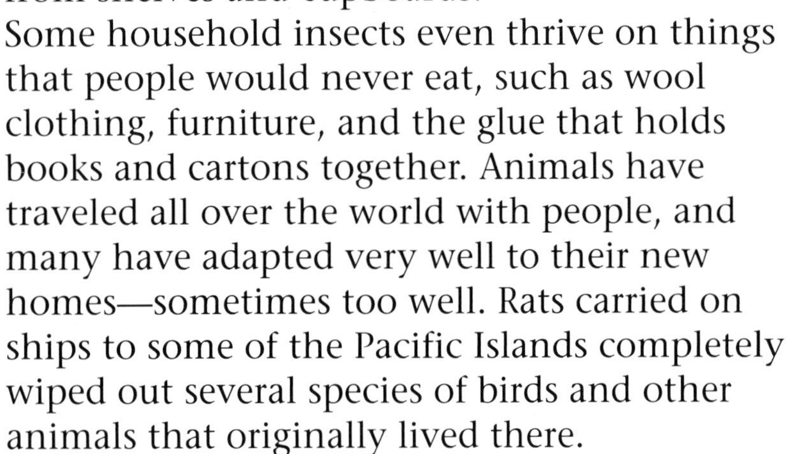

Some household insects even thrive on things that people would never eat, such as wool clothing, furniture, and the glue that holds books and cartons together. Animals have traveled all over the world with people, and many have adapted very well to their new homes—sometimes too well. Rats carried on ships to some of the Pacific Islands completely wiped out several species of birds and other animals that originally lived there.

A DANGEROUS BUZZ
Mosquitoes are among the least loved of our companions, mainly because they feed on us! Only the female mosquito's mouth is equipped with the necessary piercing and sucking tools. She has six pointed needles that fit together to form a slender tube similar to a hypodermic needle. When the mosquito bites, its saliva enters the wound immediately to prevent the blood from clotting. Not all mosquitoes attack people; most species feed on other animals, including amphibians, reptiles, and birds.

A CLOSE COMPANION
The house mouse has always lived alongside people, feeding on the things that we store or discard. It lives in houses, shops, bars, and mills, but can also survive outdoors in hedges and bushes. The mouse is a nocturnal animal, and only ventures out to feed at night. It eats stored foods, such as seeds, grain, and preserved foods in general. It spoils the stocks of food accumulated by humans with its droppings, but only eats the small amount necessary to maintain itself, which is about .12 ounces (3.5 g) a day. Outside it feeds mainly on insects and seeds and is itself eaten by owls, cats, foxes, and many other predators. Food also determines the social life of the mouse. Mice with plentiful food supplies form stable groups, while others form smaller and less stable families.

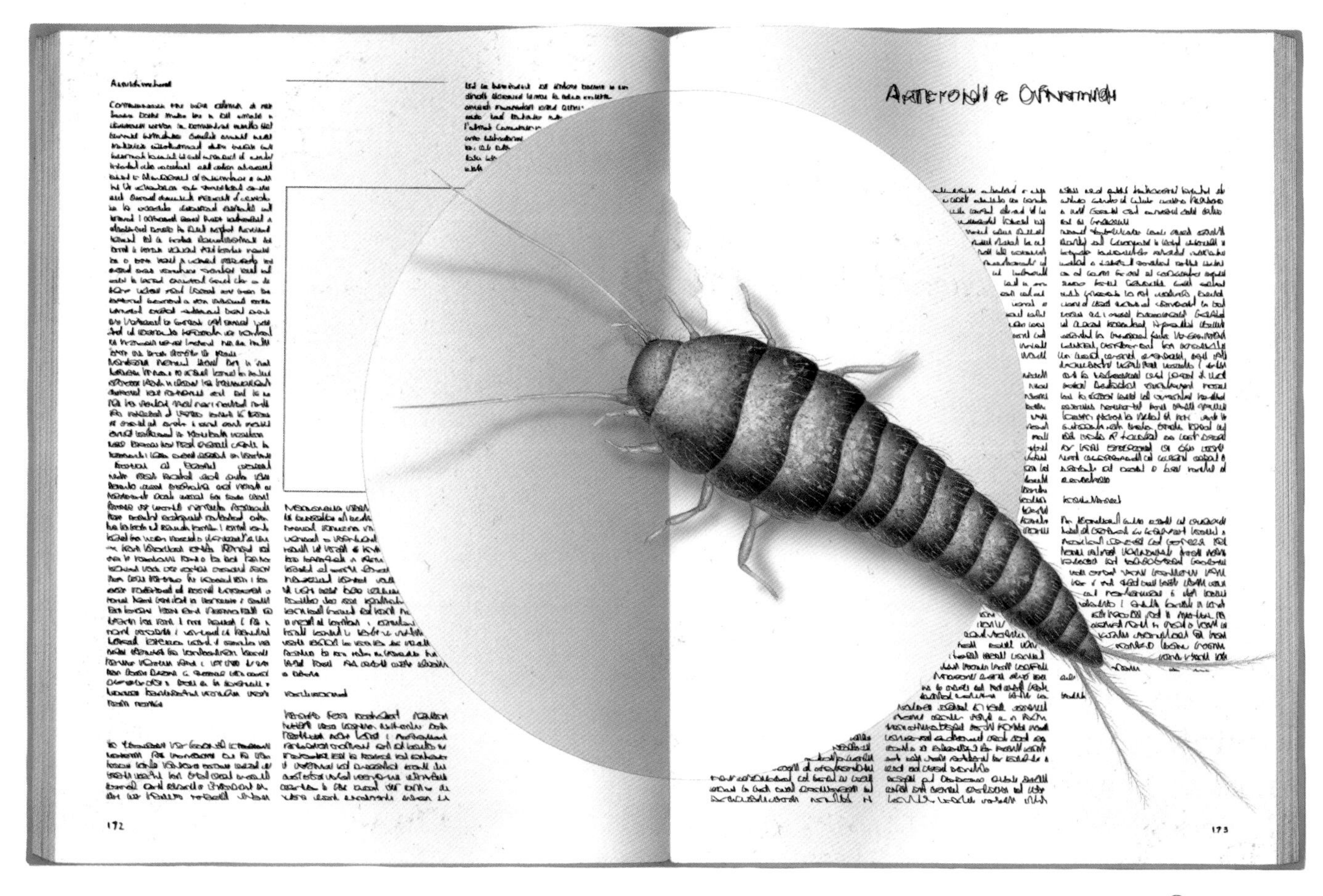

GLUE-EATERS

The silverfish is one of the simplest insects. It has no wings, and its mouth has a very basic structure. It is covered with tiny silvery scales, after which it is named. If you try to grab a silverfish, you will be left with only these shiny scales in your hand. It feeds on anything it finds in the kitchen, but prefers carbohydrates, such as sugar and flour. Silverfish are often found in books, where they feed on the glue used for binding.

GUESTS IN THE PANTRY

The cadelle beetle feeds on flour and flour-based products. It eats cereals of all types, potatoes, pasta, and nuts. It also likes fruit. Entomologists (scientists who study insects) believe that the beetle was originally a predator and that only when it came to live with people did it become a pest. Even now, it will eat the eggs and larvae of other insects whenever it can, but because some of these other insects are pests themselves, the cadelle is not all bad. In the wild, the beetle lives under bark and in rotting wood.

The Specialists

Because most animals use their jaws to gather or capture their food, it is not surprising that these are the parts that have undergone the most changes as animals have specialized in different diets. Insects show a particularly wide range of feeding equipment adapted for dealing with almost every kind of food. Butterflies and moths have tongues like drinking straws for sucking nectar from flowers: the housefly mops up liquids with a spongy pad at the end of its tongue, while bugs and mosquitoes have needlelike jaws with which they pierce plants or animals and suck out the juices. Beetles, grasshoppers, and many other insects have tough, biting jaws for chewing leaves and other solid foods. The beaks of birds and the **muzzles** of mammals can be likened to precision tools, each fashioned to carry out a particular job. The beak of a parrot or macaw, for example, can open a hard nut far more efficiently than a human-made nutcracker. Specialization may mean that an animal can feed on a particular food without competition, but too much specialization is not necessarily a good thing. If the food source becomes scarce or dies out, the specialist is doomed.

SKIMMING THE SURFACE
Skimmers, or scissorbills, are very well-named birds. Related to terns and gulls, their most striking feature is the long, scissorlike bill of which the lower part may be up to one-third as long again as the upper part. When feeding, skimmers fly over the surfaces of lakes, rivers, and lagoons with the top of the bill cutting through the water. When the lower part of the bill touches a fish, the head is pulled downward and the scissorbill snaps shut on the prey, trapping it sideways between the "scissors."

VACUUM CLEANER OF THE PAMPAS
Anteaters live in the pampas, tropical grasslands, and forests from Mexico to northern Argentina. They eat termites and ants. Their elongated skulls, tubular muzzles, and long, sticky tongues are perfect adaptations for penetrating the nests of termites and ants. The sharp, curved claws on their front paws are used to break open the insects' nests, as well as for defense. Armed with this digging and sucking equipment, anteaters can consume more than 30,000 insects a day. Dense fur protects the anteaters from angry ants and termites.

MORE ABOUT GLOWWORMS
The New Zealand glowworm is the larva of a small fly that lives in the temperate rain forests. The female flies deposit their eggs on the walls of the Waitomo Caves in the North Island of New Zealand (where the glowing worms have become a tourist attraction). When the larvae leave the eggs, they excrete long, sticky, luminous silk threads that attract and trap other tiny insects that inhabit the caves. After 6 to 8 months, depending on the availability of food, the larvae turn into adult flies.

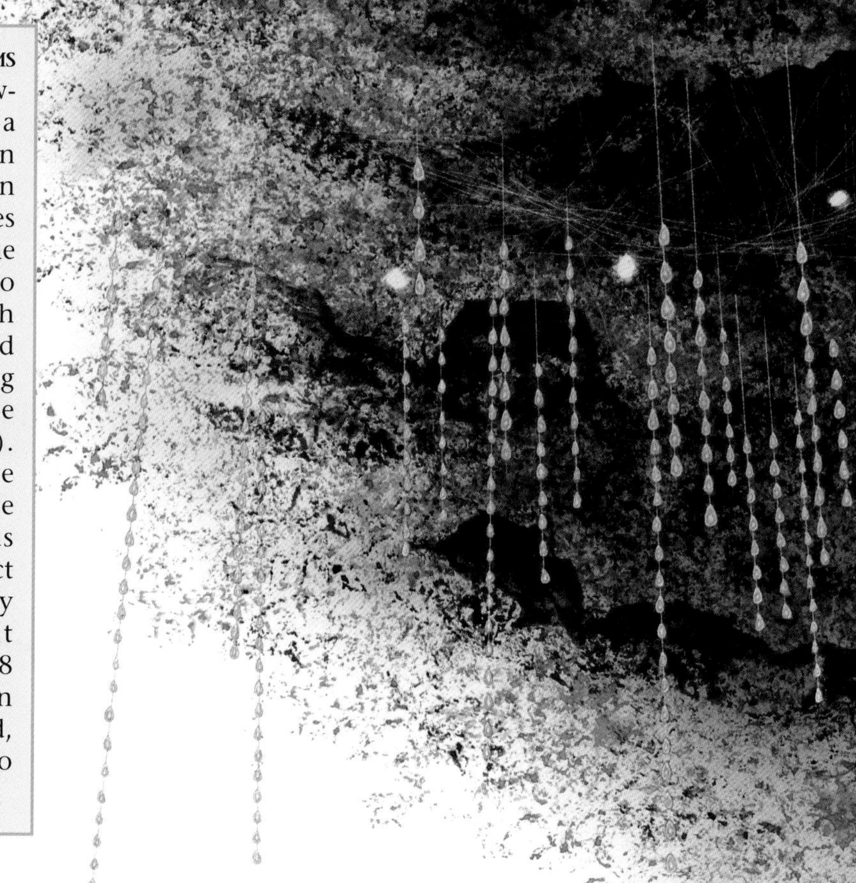

STICKY FISHING LINES

The larvae of a fly that lives in caves in Australia and New Zealand use light when hunting. Although called a glowworm, it is not related to the European glowworm, which is a beetle. Each glowworm produces a long, sticky thread that hangs down toward the floor. The threads are hung close together, creating a net that small insects fly into. To attract the insects, the worms glow slightly, lighting the caves with a pale green color. The light is reflected by tiny globules of sticky substance in the threads. The worms gather the insects caught on the threads and eat them.

SURE-SHOT TONGUE

Chameleons are precision hunters. Their whole body and behavior are directed toward enabling their tongues to strike as rapidly and precisely as possible. In some species the tongue can extend to almost twice the length of the body. It has a sticky patch on the end so that, once hit, the prey cannot escape. The chameleon relies on its eyesight to find prey and to take aim. Its eyes are situated on swiveling **turrets**, each one independent from the other. When the chameleon is on the lookout for prey, its field of vision is quite wide. Once the prey has been spotted, its eyes turn to the front of the head, and it uses binocular vision to shoot its tongue at the unsuspecting insect.

Digesting Food

The quest for nourishment does not end when food has been gathered or caught. It continues inside the animal's body with the process of digestion, during which the food is broken down by digestive juices and converted into sugars and other simple substances. These are carried by blood to the cells and tissues of the body, where they are combined with oxygen to release the energy that the body needs. Plant matter is harder to digest than animal flesh, so herbivores usually have longer digestive canals than carnivores. The digestive canals of most herbivorous mammals also contain bacteria that aid digestion. Cattle, deer, and antelope have two chances to digest their food. They bring it back up into their mouths and chew it a second time, which is called "chewing the cud." Rabbits and hares even eat their own droppings so that their food makes two complete journeys through the digestive system. Spiders and some other predators start the digestion even before they swallow their prey, by injecting digestive juices.

BORN TO EAT
Butterflies are only the last stage in the life cycle of these insects. The lengthiest period of the insect's life is often the caterpillar stage. Caterpillars usually have much larger digestive systems than adult insects and spend most of their lives eating plants. In this way they accumulate large stores of food, which they use to transform themselves into adults. Because their main job is to eat, caterpillars don't have reproductive systems. Reproduction is left to adult insects.

TOO HEAVY TO FLY
The hoatzin is a strange bird that lives in the tropical forests of South America. The hoatzin is one of very few birds that feeds exclusively on leaves, which are generally quite difficult to digest. The birds have to swallow large quantities of leaves in order to get enough useful food from them. Their digestive systems therefore are bulky, leaving little room for the flight muscles. With its large stomach and small muscles, the hoatzin has difficulty flying and never flies very far.

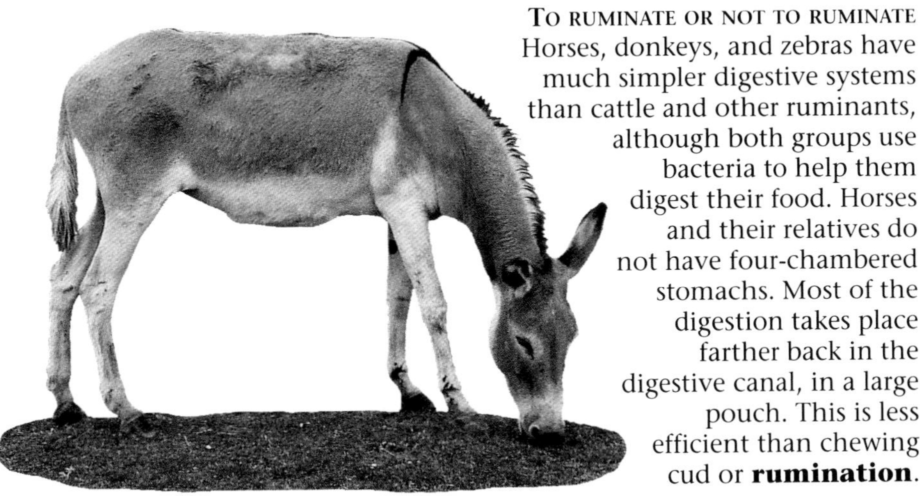

To ruminate or not to ruminate
Horses, donkeys, and zebras have much simpler digestive systems than cattle and other ruminants, although both groups use bacteria to help them digest their food. Horses and their relatives do not have four-chambered stomachs. Most of the digestion takes place farther back in the digestive canal, in a large pouch. This is less efficient than chewing cud or **rumination**.

More about digestion
Digestion usually begins in the mouth, where food is chewed and mixed with saliva. Human saliva contains a substance called ptyalin that reacts with starchy foods and begins to convert them to sugar. Other digestive juices produced in the stomach and intestines break down proteins and fats. Each kind of animal produces its own kinds of digestive juices, designed to deal with its diet. Meat-eaters do not need much ptyalin in their saliva, because their food does not contain much starch.

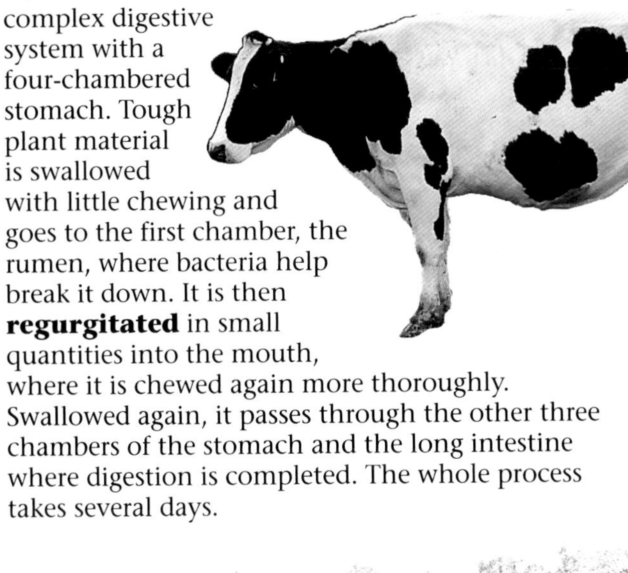

Cattle and other ruminants have a complex digestive system with a four-chambered stomach. Tough plant material is swallowed with little chewing and goes to the first chamber, the rumen, where bacteria help break it down. It is then **regurgitated** in small quantities into the mouth, where it is chewed again more thoroughly. Swallowed again, it passes through the other three chambers of the stomach and the long intestine where digestion is completed. The whole process takes several days.

Deadly saliva
Poisonous snakes use venom to kill their prey. Once the prey is immobilized or dead, they can eat it in peace. The glands that produce the venom are similar to saliva glands, and the poison itself is made up in part of enzymes like those in human saliva or the digestive system. Once the venom penetrates the victim's body, it works in two ways. First it immobilizes the prey (using different mechanisms according to the species of snake); then it slowly destroys the victim's body tissue. The enzymes in the venom are not sufficient to completely digest the prey, but they do help break down the food later when the snake has swallowed it.

Living Food Stores

Many animals build up stores of food in their bodies in preparation for times of scarcity. This is a good way of ensuring that the animals have enough food to tide them over periods of bad weather, but the reserves must not slow the animals down and make them more obvious to predators. It is mainly animals that live in deserts or on dry grasslands that store food in their bodies. They build up their reserves before the dry season, when food is often hard to find. Several rodents and lizards store food in their tails. Mammals that sleep through the winter, such as bears, dormice, and marmots, build themselves up by stuffing themselves with nuts and other high-energy foods in the fall. Although their bodily activities slow down during the winter, they still need some energy to stay alive. Whales deposit thick layers of blubber under their skins while feeding in the Arctic or Antarctic oceans in the summer. This fat, which may account for over half their weight, keeps them going when they migrate to tropical areas to breed in the winter.

MORE ABOUT ANTS

There are more than 20,000 species of ants. Most of them live in the warmer regions. Ants range in size from about 1/8 of an inch to 1 inch (.2 to 2.5 cm) and are usually brown, yellow, red, or black in color. They are social insects, living together in well-organized colonies. There are normally three **castes** within the colony: queens, males, and workers. Sometimes the worker caste is divided into more than one group, depending on the activities undertaken. The queens spend their lives laying eggs. Males mate with the queens to fertilize their eggs. The workers are all females. They take care of the nest, protect it from invaders, collect all the food, and they feed and look after the young ants. Some species of ants even enslave members of other species, forcing them to work for them.

STOCKING UP FOR WINTER

Bears living in cold climates sleep for long periods in the winter months. Before the bad weather starts, they eat a large quantity of all sorts of food, from berries and seeds to small mammals. In this way they build up a reserve of fat under their skins that they slowly absorb during the winter months as they sleep. Strictly speaking, bears and other large carnivores cannot be said to hibernate because their metabolism, or bodily activity, only slows by half: in dormice and other small mammals it drops to almost zero. The bears go to sleep for much of the time, because there are not many juicy shoots and leaves for them to eat in the winter. Babies are usually born in the winter, and their mothers keep them warm in their sleeping dens.

BOTTLE-FEEDING
Social insects are famous for the way they divide labor among the members of their group. In some species of ants, the degree of specialization is extreme. Some Australian ants feed on the nectar of desert flowers. The nectar is only available at certain times of the year and must be stored for use during periods of scarcity. The ants collect the nectar and give it to another caste of workers, who store it in their hugely enlarged abdomens. These workers act as living storerooms; they hang stomach-down in special areas of the ant nest. When nectar is scarce, the bottle workers feed all the other members of the colony from their stores.

A WATER RESERVE
The camel's ability to survive in the dry heat and cold of the desert depends on many external and internal adaptations. Thick eyelashes protect the eyes from wind-blown sand, and the animals can also close their nostrils to keep it out. Their wide hooves help them to walk over the soft sand without sinking. But internal mechanisms, including the hump, are perhaps the most impressive. The hump, for example, acts as a store for fat, which can be transformed into water by means of special metabolic processes. With this reserve on their backs, camels can roam far from sources of water, such as oases and human settlements.

Glossary

Adaptation Evolutionary change in a plant or animal that increases its ability to survive and reproduce in its particular environment.

Amphibian Member of the vertebrate class Amphibia, which lives both on land and in water.

Baleen Plates of a material called keratin that hang down from the upper jaw of toothless whales. The plates act like sieves to trap food from the water.

Camouflage Means by which animals blend into their surroundings or otherwise deceive predators and escape their attention.

Carrion Dead flesh.

Caste A "social class" among some social insects, such as ants and termites.

Cecum A pouch open at one end in which the large instestine begins.

Cellulose A carbohydrate that forms the framework of plant cell walls.

Crustacean Member of a group of hard-shelled, mainly aquatic arthropods, including crabs, shrimp, and barnacles.

Deciduous Describes trees that shed their leaves each year at the end of the growing season, usually in the fall.

Echinoderm Any of about 6,000 species of marine invertebrates with spiny skins, including starfish and sea urchins.

Echolocation Finding the position of an object by sending out high-pitched sounds and measuring the time taken for echoes to return from it.

Ecosystem An ecological unit such as a pond, forest, or desert, consisting of different plant and animal communities, as well as the nonliving environment.

Evolution Process by which plants and animals change over successive generations, often resulting in better adaptation to the environment and eventually producing new species.

Extinction Death of the last remaining individuals of a species of plant or animal.

Food chain Series of organisms, each one of which is a food source for another. A simple chain might start with grass, which is eaten by rabbits, which are eaten by foxes or hawks. The remains of dead plants and animals may be eaten by scavengers or broken down by bacteria.

Gland Organ in an animal's body that secretes useful chemical substances, such as hormones and digestive juices.

Harem Permanent or temporary group of animals, consisting of one breeding male and several females.

Hatchling An animal that has recently hatched from an egg.

Hibernation Period of winter sleep, during which an animal's metabolism slows down and its temperature drops.

Insectivore An animal that feeds on insects.

Invertebrate Any animal without a backbone.

Krill Small, shrimplike crustaceans that live in huge swarms in the oceans and are an important food source for baleen whales and other sea creatures.

Larva (plural larvae) Stage in the metamorphosis of certain life forms, such as butterflies and frogs.

Mammal Any member of the class Mammalia. Most have hair and give birth to active, live young. The female feeds her young on milk from her body.

Marsupial Mammals, such as kangaroos, opossums, and koalas, that give birth to tiny young that continue to develop in a pouch on their mother's belly.

Metabolism General term for all the chemical processes that occur in living things, involving, for example, growth, breathing, and energy production.

Microscopic Too small to be seen without the aid of a microscope.

Migration Regular movement of animals from one area to another and back again at certain times of the year.

Mollusk Any member of the phylum Mollusca; most mollusks are soft-bodied, though many have shells.

Muzzle Projecting part of a mammal's head, including its nose and mouth.

Nocturnal Describes animals that are active during the night.

Omnivore Any animal that feeds on both plants and other animals.

Opportunistic feeder Animal that will eat whatever it can find, by hunting, scavenging, or stealing food from other animals.

Parasite Plant or animal that lives in or on another (the host), and feeds on it.

Phloem Plant tissue in which the sugars and other food materials are carried from one part of the plant to another.

Photosynthesis The process by which green plants use sunlight, water, and carbon dioxide to produce food.

Placenta An organ that forms in the uterus of most female mammals during pregnancy. It links the growing baby to the mother, supplying it with oxygen and nutrients and removing wastes.

Plankton Minute plants and animals that float on or just beneath the surface of lakes, rivers, and oceans, and provide valuable food for larger animals.

Pollen Dustlike grains containing male sex cells in seed-bearing plants.

Pollination Process by which the wind or animals carry pollen from the male organs to the female organs of seed-bearing plants.

Predator Any carnivorous animal that hunts and kills other animals (prey).

Prothorax First segment of an insect's thorax (the part that lies between the head and abdomen).

Protozoan An animal-like single-celled organism.

Regurgitate To bring swallowed, partially digested food up again.

Reptile Any member of the class Reptilia; vertebrates with scaly skin.

Rodents Large group of mammals that includes rats, mice, and beavers.

Rumination The digestive process of cattle, sheep, and other grazing mammals. Ruminants have special multichambered stomachs from which partially digested food is regurgitated into the mouth, chewed, and swallowed again for complete digestion.

Scavenger An animal that feeds on dead bodies and other decaying matter.

Symbiosis Close association between individuals of different species, from which both benefit.

Talons Elongated, hooked claws of birds of prey.

Tannins Bitter substances in plants.

Territory Area inhabited and defended by an animal or group of animals against others of the same species.

Tropics Regions lying on each side of the equator between the Tropic of Cancer and the Tropic of Capricorn.

Turrets Devices that hold several lenses.

Venom Poison, especially a poison that is injected into the prey or an enemy by a bite or a sting.

Vertebrates Animals with a backbone.

Index

Further Reading

Facklam, Howard and Margery. *Parasites*. New York, NY: 21st Century, 1994.
Kite, L. Patricia. *Blood-Feeding Bugs & Beasts*. Millbrook, 1996.
Lovett, Sarah. *Extremely Weird Animal Hunters*. Davidson, 1997.
Perry, Phyllis J. *Armor to Venom: Animal Defenses*. Watts, 1997.
Simon, Seymour. *Ride the Wind: Airborne Journeys of Animals and Plants*. Harcourt Brace, 1997.

Acknowledgments

The Publishers would like to thank the following photographers and archives for permission to reproduce pictures and for their assistance in providing pictures. The sources of the photographs are listed below. The following short terms have been used:

CAPPELLI = Giuliano Cappelli, Florence
CERFOLLI = Fulvio Cerfolli, Rome
JACANA = Jacana, Paris
NARDI = Marco Nardi, Florence
OKAPIA = Okapia, Frankfurt
OSF = Oxford Scientific Films, London
OVERSEAS = Overseas, Milan
PANDA = Panda Photo, Rome

Front cover: **TL** V. Giannotti/PANDA; **CL** F. Pölking/OVERSEAS; **BL** S. Dalton/OSF-OVERSEAS; **R** CAPPELLI; **Back cover:** CAPPELLI; **1** N. Rosing/OVERSEAS; **4** CAPPELLI; **5** NARDI; **6C** Laboute/JACANA-OVERSEAS; **6B** NARDI; **7** K. Senani/OSF-OVERSEAS; **8B** J. Foott/PANDA; **8T** CAPPELLI; **9T** A. Petretti/PANDA; **9B** CAPPELLI; **10B** C. Galasso/PANDA; **10T** OSF-OVERSEAS; **11T** CAPPELLI; **12T** V. Gianotti/PANDA; **12BR** CAPPELLI; **13B** R. Tyrrell/OSF-OVERSEAS; **14C** A. Nardi/PANDA; **14B** F. Roberdeau/JACANA-OVERSEAS; **15T** Gavazzi/OVERSEAS; **15CL** P. Pilloud/JACANA-OVERSEAS; **15CR** A. Petretti/PANDA; **16T** B. Cranston/PANDA; **16B** M. Wendler/PANDA; **17T** G. Bernard/OSF-OVERSEAS; **17B** N. Rosing/OVERSEAS; **18C** M. Fogden/OSF; **19T** C. Bagnoli/PANDA; **19B** M. Stouffer/OSF-OVERSEAS; **20B** G. Pollini/PANDA; **20T** G. Marcoaldi/PANDA; **21T** J. Foott/PANDA; **22T** Gerard/JACANA-OVERSEAS; **23B** De Roy/OVERSEAS; **24T** Z. Leszczynski/OVERSEAS; **24B** D. Salussoglia/PANDA; **25T** G. Picchetti/PANDA; **25B** T. Saginata/OVERSEAS; **26T** S. Maslowski/PANDA; **26B** CAPPELLI; **27T** F. Camara/PANDA; **27B** CAPPELLI; **28T** VARIN-OVERSEAS; **29T** C.Dani/OVERSEAS; **30T** M. Harvey/PANDA; **30B** F. Ehrenstrom/OSF-OVERSEAS; **31T** F. Andreone/OVERSEAS; **32T** J. Foott/PANDA; **32C** S. Montanari/PANDA; **32B** F. Di Domenico/PANDA; **33C** PANDA; **33B** D. P. Wilson/PANDA; **34T** J. Philippe/VARIN-OVERSEAS; **34B** G. Bernard/OVERSEAS; **35T** F. Pölking/OVERSEAS; **36T** F. Pölking/OVERSEAS; **37T** CAPPELLI; **37C** JACANA-OVERSEAS; **38T** E. Dragesco/PANDA; **38B** R. Mayr/OSF; **39B** M. Bonora/PANDA; **40T** A. Nardi/PANDA; **40B** CERFOLLI; **41T** C. Jones/OVERSEAS; **41C** A. Nardi/PANDA; **41B** N. Dennis/PANDA; **43T** J. Nest/OVERSEAS; **43C** E. Coppola/PANDA; **44T** CAPPELLI; **45C** CAPPELLI; **45T** F. Bruemmer/PANDA; **46C** A. Root/OKAPIA; **47TL** G. Bernard/OSF; **47TR** J. Foott/PANDA; **48T** E. Stella/PANDA; **48B** OSF-OVERSEAS; **49T** OVERSEAS; **49BR** Leszczynski/OVERSEAS; **50TL** F. Pölking/OVERSEAS; **51B** E. Dragesco/PANDA; **52T** G. Bernard/OSF-OVERSEAS; **52B** R. Tyrrell/OSF-OVERSEAS; **53B** E. Stella/PANDA; **54B** VARIN-JACANA-OVERSEAS; **54T** K. Sandved/OSF-OVERSEAS; **55B** S. Downer/OSF-OVERSEAS; **56T** R. Savelli/PANDA; **56B** J. Munoz/PANDA; **57T** NARDI; **57C** NARDI; **58B** CAPPELLI; **59B** CAPPELLI